Distributed Large-Scale Dimensional Metrology

Fiorenzo Franceschini · Maurizio Galetto
Domenico Maisano · Luca Mastrogiacomo
Barbara Pralio

Distributed Large-Scale Dimensional Metrology

New Insights

Springer

Prof. Fiorenzo Franceschini
Department of Production Systems and
Business Economics (DISPEA)
Politecnico di Torino
Corso Duca degli Abruzzi 24
10129 Turin, Italy
e-mail: fiorenzo.franceschini@polito.it

Prof. Maurizio Galetto
Department of Production Systems and
Business Economics (DISPEA)
Politecnico di Torino
Corso Duca degli Abruzzi 24
10129 Turin, Italy
e-mail: maurizio.galetto@polito.it

Dr. Domenico Maisano
Department of Production Systems and
Business Economics (DISPEA)
Politecnico di Torino
Corso Duca degli Abruzzi 24
10129 Turin, Italy
e-mail: domenico.maisano@polito.it

Dr. Luca Mastrogiacomo
Department of Production Systems and
Business Economics (DISPEA)
Politecnico di Torino
Corso Duca degli Abruzzi 24
10129 Turin, Italy
e-mail: luca.mastrogiacomo@polito.it

Dr. Barbara Pralio
Department of Production Systems and
Business Economics (DISPEA)
Politecnico di Torino
Corso Duca degli Abruzzi 24
10129 Turin, Italy
e-mail: barbara.pralio@polito.it

ISBN 978-0-85729-542-2 e-ISBN 978-0-85729-543-9

DOI 10.1007/978-0-85729-543-9

Springer London Dordrecht Heidelberg New York

British Library Cataloguing in Publication Data
A catalogue record for this book is available from the British Library

Cover design: eStudio Calamar S.L.

Printed on acid-free paper

Springer is part of Springer Science+Business Media (www.springer.com)

Foreword

Large-scale metrology, a remarkable addition to the field of dimensional measurement, was prompted in the last decades mainly by the pressing needs of several strategic industrial sectors, leading to a dynamic expansion of both types and capabilities of methods exploited. The challenge of rapid, accurate measurement of coordinates pertaining to objects whose size may well exceed that of a football field is indeed a formidable one, even more considering the severe ambient conditions prevailing, for example, in shipyards.

Distributed systems, catering for rapid data acquisition at a number of locations via mobile spatial coordinate measurement devices, address requirements hard to meet with traditional, time consuming methods, exploiting e.g., surveying and topographic techniques. How critical is accurate, timely geometric description of large ship subassemblies was shown by catastrophic failures at sea, traced back to abusive jacking resorted to as last resort at assembly to compensate mismatch detected too late. Some embarrassing delays in aircraft production did also occur, owing to late appreciation of fatal glitches and misunderstandings regarding spatial measurement of components.

Development of rugged, portable devices providing fast coordinate acquisition under control of a dedicated instrument systems goes a long way toward meeting industrial requirements. Several solutions proposed recently are described in this book, and their main features discussed, as well as a comprehensive survey of what has been achieved in this area by leading research laboratories, together with results obtained by a dedicated research group, whose successful development we had the privilege of witnessing right from the beginning.

A detailed description of operating principles, and performance characteristics, is presented, along with a thorough treatment of such key aspects as system calibration, self-diagnostics and evaluation of measurement uncertainty. Besides state of art description, practical guidance to users is provided, a most welcome feature for operators involved in actual operation. We wish to congratulate the authors for sharing with all those concerned their outstanding experience, and packing so much valuable information in this book.

Raffaello Levi
Professor Emeritus at the Fourth School of Engineering of
Politecnico di Torino, Italy

Sergio Rossetto
Dean of the Fourth School of Engineering of
Politecnico di Torino, Italy

Preface

The last decades have shown great improvements in the field of engineering dimensional metrology. The introduction of modern computational systems, ever more compact microelectronic devices and innovative materials have given a significant impulse to the accuracy and application of traditional devices. Recent developments have included applications both for extremely small-sized and for extremely large-sized objects. Following this, two new branches of dimensional metrology have demonstrated their maximal potentiality: "nano-metrology", related to the measurement of extremely small objects, and "large-scale dimensional metrology" (LSM), which refers to the measurement of large engineering structures.

To begin with nano-metrology, the current widespread success can be ascribed to the new generations of scanning electron microscopes (SEMs) and, even more, to atomic force microscopes (AFMs), which allow us to obtain precise and accurate measurements of very small elements and structures. On the other hand, large-scale dimensional metrology has reached its historic turning point with the introduction of modern very accurate and flexible optical systems.

The aim of the present book is to trace and discuss the most recent developments in the field of LSM, with particular attention to so-called distributed systems. Special emphasis is given to topics concerning sensor positioning, system calibration and self-diagnostics, which represent the key elements for the correct functioning of these systems.

This book is directed at academics, practitioners and consultants who are responsible for designing the best solutions and tools to manage LSM processes. The text materials can be used in programs for professional practitioners, including quality engineers, manufacturing engineers and procurement specialists. Professionals may also use the material for self study.

This book is organized in eight chapters. Chapter 1 introduces large-scale metrology, discussing its goal and giving a short summary of classical instruments. Some of the most recent instruments are then introduced providing a structured classification.

Chapter 2 describes the indoor-GPS (iGPS^{TM}), which currently represents a turning point for LSM distributed systems.

Chapter 3 is dedicated to the technical description of Mobile Spatial coordinate Measuring System (MScMS), a prototypal system developed from the research activity carried out at the Industrial Metrology and Quality Engineering Laboratory of Department of Production Systems and Business Economics (DISPEA) at the Politecnico di Torino. The UltraSound (US) version (MScMS-I) and the InfraRed (IR) version (MScMS-II) are described in detail. A general comparison between these two versions and other existing systems is also presented.

Chapter 4 deals with network device positioning and coverage of the measuring area, while Chaps. 5, 6 and 7, respectively approach the topics of system calibration, system on-line self-diagnostics and improvement of system metrological performance. Uncertainty budget evaluation is treated in Chap. 8.

All these chapters face the respective topics from a general point of view, enriching the explanation by introducing some practical examples.

Turin, October 2010

Fiorenzo Franceschini
Maurizio Galetto
Domenico Maisano
Luca Mastrogiacomo
Barbara Pralio

Acknowledgments

This book and the whole research program "Large-scale coordinate metrology: study and realization of an innovative system based on a network of distributed and cooperative wireless sensors" (PRIN 2008 prot. no. 200853ZT3Z) have been financially supported by the Italian Ministry of University and Scientific and Technological Research (MIUR).

Authors wish to gratefully acknowledge the support of all colleagues and friends who have contributed to the realization of the book with stimulating suggestions and helpful comments.

Contents

Acronyms and Abbreviations

2D	Two-dimensional
3D	Three-dimensional
ADM	Absolute distance measure
AFM	Atomic force microscope
AGV	Autonomous guided vehicle
AoA	Angle of arrival
CCD	Charge-coupled device
CMM	Coordinate measuring machine
CNC	Computer numerical control
CSAIL	Computer science and artificial intelligence lab
CWA	Conventional weight aggregation
DGPS	Differential global positioning system
DISPEA	Dipartimento di Sistemi di Produzione ed Economia dell'Azienda
DOE	Design of experiments
DOF	Degree of freedom
DOP	Dilution of precision
DPU	Data processing unit
EEPROM	Electrically erasable programmable read-only memory
EF	Error function
FF	Fitness function
FOV	Field of view
GA	Genetic algorithm
GPS	Global position system
GUI	Graphical user interface
GUM	Guide to the expression of uncertainty in measurement
ICT	Information and communications technology
IEC	International Electrotechnical Commission
IEEE	Institute of Electrical and Electronics Engineers
iGPS™	indoor GPS
IR	Infrared
ISO	International Organization for Standardization

LAL	Lower acceptance limit
LED	Light emitting diode
LSM	Large-scale metrology
MIT	Massachusetts Institute of Technology
MLPU	Multivariate law of propagation of uncertainty
MOP	Multi-objective optimization problem
MPE	Maximum permitted error
MScMS	Mobile spatial coordinate measuring system
NP	Nondeterministic polynomial time
NPL	National physics laboratory
PC	Personal computer
PDOP	Position dilution of precision
PVDF	Polyvinylidene fluoride
RAM	Random-access memory
RF	Radio frequency
RH	Relative humidity
RLS	Robust least squares
ROM	Read-only memory
RSS	Received-signal-strength
SA	Simulated annealing
SEM	Scanning electron microscope
TCP	Tool centre point
TCP/IP	Transmission control protocol/internet protocol
TDoA	Time difference of arrival
ToA	Time of arrival
TOF	Time of flight
TS	Tabu search
UAL	Upper acceptance limit
UDP	User datagram protocol
US	Ultrasonic
V-bar	Vector bar (iGPS™ portable probe)
VIM	International vocabulary of basic and general terms in metrology
WSN	Wireless sensor network

Chapter 1
Large-Scale Dimensional Metrology: The New Paradigm of Distributed Systems

1.1 Goal and History

The first definition of "Large-Scale Dimensional Metrology" dates back to the early 1960s, when Berry (1961)—in his report about the techniques for precise measuring involved in the construction of "Nimrod", the 7 GeV Proton Synchrotron at the Rutherford High Energy Laboratory in Harwell (England)—referred to it as "… a field of development in which the hitherto separate skills of the surveyor and the engineering metrologist have been brought together in what is becoming known as 'Large-Scale Metrology'. Large-Scale Metrology means the measurement of dimensions usually undertaken by surveyors to limits normally associated with the workshop". Perhaps this definition was not yet completely exhaustive and fully comprehensive yet, but it contained *in a nutshell* a realistic description of what Large-Scale Dimensional Metrology was to become in a few years.

A more rigorous definition was then proposed by Puttock (1978) 17 years later in his famous paper titled "Large-Scale Metrology". According to Puttock, "The field of Large-Scale Engineering Metrology can be defined as the metrology of large machines and structures. The boundaries of this field are laboratory measurements at one end and surveying at the other. Neither boundary is well defined and … will be generally confined to the metrology of objects in which the linear dimensions range from tens to hundreds of metres".

Puttock did not restrict his definition to the mere field of application, he also included the environmental conditions and the pursued goals of the measurement as basic elements of Large-Scale Dimensional Metrology applications. He observed that Large-Scale measurements can very rarely be done in a reasonably well controlled environment. Usually, the metrologist is forced to adapt and modify equipment and techniques to suit each differing set of circumstances and to achieve the required accuracy.

F. Franceschini et al., *Distributed Large-Scale Dimensional Metrology*,
DOI: 10.1007/978-0-85729-543-9_1, © Springer-Verlag London Limited 2011

Traditionally, Large-Scale Dimensional Metrology concerns measurements of length and angle, however methods and techniques have undergone radical improvements over the years. For example, for length measurements, the so-called "classical approach" was based on the use of surveyors' tapes and wires. Best results required considerable skill and attention to detail in both equipment and techniques. A great advance was made with the introduction of optical methods. First, the advent of laser interferometry, and then the use of "range finders", devices involving the time-of-flight of an optical or microwave pulse, marked a fundamental turning point in the way towards the modern conception of Large-Scale Dimensional Metrology (Froome and Bradsell 1966).

In angle measurement, given that the angular relationship of features of large engineering structures depends very much on the detail of the particular structure, in general the most utilized methods have always been optical. Metrologists have advised particular attention towards the limitations imposed by the propagation of light through the atmosphere right from the first applications (Puttock 1978; Estler et al. 2002). As a matter of fact, the propagation of light rays may be affected by errors induced by refraction effects or variation of speed of light during their propagation in the atmosphere.

Recent years have shown significant advances across a broad range of technologies, including laser interferometry, absolute distance metrology, the development of powerful diode lasers and very high density CCD cameras, and the continuing evolution of very powerful general purpose computers and associated software. The last ones have greatly improved the speed of data acquisition and elaboration towards levels which, only 20 years ago, were unimaginable, as well as giving elevated accuracy and system flexibility. Nowadays the development and refinement of optical systems has led to a widespread use of instruments, ranging from theodolites, laser-trackers, laser-radars, iGPS™ to target-less scanning systems using absolute distance ranging and powerful digital photogrammetric methods.

Historically, typical examples of application of Large-Scale Dimensional Metrology refer to the measurement of large structures such as modern radio telescopes in astronomical observatories (Puttock 1978; Estler et al. 2002), trusses in spatial stations, liquefied natural gas tanks; alignments or deformations such as in the case of deformation monitoring of ore crushers, straightness errors in large machine tools, or alignment of large particle accelerators; large manufacture assembly such as ships or aircraft fuselage and wings (Estler et al. 2002).

Recently, a new generation of Large-Scale Dimensional Metrology systems has established itself, the so-called "distributed systems". These generally consist of a set of remotely distributed sensing devices coordinated by one or more central processing units, which is in charge of data acquisition and post-processing elaboration to provide measurement results. The distributed approach can be limited to the spatial location of the devices, which are simply remote sensing units, providing reference points in three-dimensional space, or it can be enlarged by attributing the role of intelligent agents to remote sensing devices, i.e., autonomous entities, which cooperate and coordinate their activities to achieve the common objective of performing the measurement. Many actual systems and research prototypes are moving in that direction.

The paradigm of Large-Scale Dimensional Metrology, which consists in transferring the measuring system to the measured object place rather than vice versa, actually finds its effective full application in distributed systems. The goal is to reconstruct the "measuring machine" all around the measured object, and to organize the positioning of the sensing devices in order to optimally cover the measuring volume. If the system is based on "intelligent" devices, they should also be able to reorganize themselves with the aim of improving or preserving a required level of quality of measurement.

Many topics about this specific approach to Large-Scale Dimensional Metrology are still the subject of discussion within the scientific community. Some issues still need to be studied further.

Much current research focuses on accuracy improvement and the evaluation of measurement uncertainty. A family of international standards for a univocal approach to performance evaluation of measuring systems and uncertainty evaluation is still lacking. Some approaches in this direction are based on existing standards for Coordinate Measuring Machines (CMMs) and optical systems. At present, some efforts towards the establishment of specific standards for those systems which have reached a consolidated technological level and significant commercial diffusion are in the course of development. For example, Part 10 of ISO 10360 will be devoted to the performance evaluation of laser trackers (Peggs et al. 2009).

Economic impact is a crucial topic for the application of Large-Scale Dimensional Metrology systems. From a technical point of view, thanks to optical approaches, ambitious levels of accuracy can be reached (see, for example, laser-radars and iGPS™), nevertheless costs still remain a problem (Galetto et al. 2011). On the other hand, in many applications, acceptable metrological performance can be obtained with economically affordable systems. Today's challenge is to design and produce low-cost, portable and user-friendly systems, with adequate level of accuracy for Large-Scale applications.

Considering that for most techniques used in Large-Scale Dimensional Metrology, the set-up is very flexible in comparison to CMMs or similar machines, the relative position of the various components of the measuring system have a fundamental effect on system performance and the estimation of uncertainty. Great strides have been made in trying to provide the same infrastructure of standards for these flexible measuring systems, unfortunately these advances have produced a number of regional standards, but an international reference is yet to come.

The scope of the present book is to give a picture of the state-of-the-art of distributed systems for Large-Scale Dimensional Metrology applications, focusing attention on the most recent research. Topics concerning sensor positioning, system calibration and self-diagnostics, which represent the key elements for the correct functioning of these systems, are also discussed in detail.

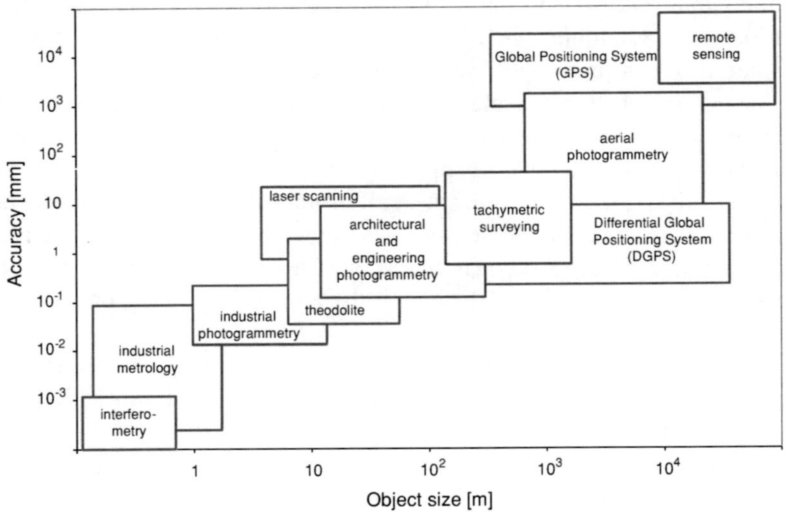

Fig. 1.1 Relationship between object size and accuracy for different measurement methods (Luhmann et al. 2006). (with permission)

1.2 Challenges of Large-Scale Dimensional Metrology

The primary purpose of Large-Scale Dimensional Metrology is the three-dimensional measurement of large-sized objects. Traditionally, there are strong links between Large-Scale Dimensional Metrology and the techniques of surveying, particularly in the areas of adjustment methods and engineering surveying. With the increasing application of Large-Scale Dimensional Metrology approaches to industrial manufacturing and quality control, links have been created in many directions.

Figure 1.1 gives a view of the relationship between size of measured object, guaranteed measurement accuracy and relevant technology. Although there is no hard and fast definition, it may be said that Large-Scale Dimensional Metrology accuracy refers to objects ranging from 10 to 1,000 m in size, with accuracies around 0.1 mm at the smaller end and 10 cm at the larger end (for large structures in space and construction industries).

Before optical methods were introduced, lengths and angles were measured by opportunely shaped tapes, levels, sights and graduated staffs. These techniques were employed mostly for surveying, but also for engineering dimensional measurements for large structures and objects. Referring, for example, to tapes, millimetre accuracy could be achieved by applying proper corrections for standard, catenary, tension, temperature and slope to the measured distance. The introduction of optical methods, firstly for the measurement of angles and then for large distances, has determined wide enhancement in terms of accuracy and ease of use.

Optical methods, which use light as information carrier, lie at the heart of non-contact 3D measurement techniques. They can be classified as follows (Luhmann et al. 2006):

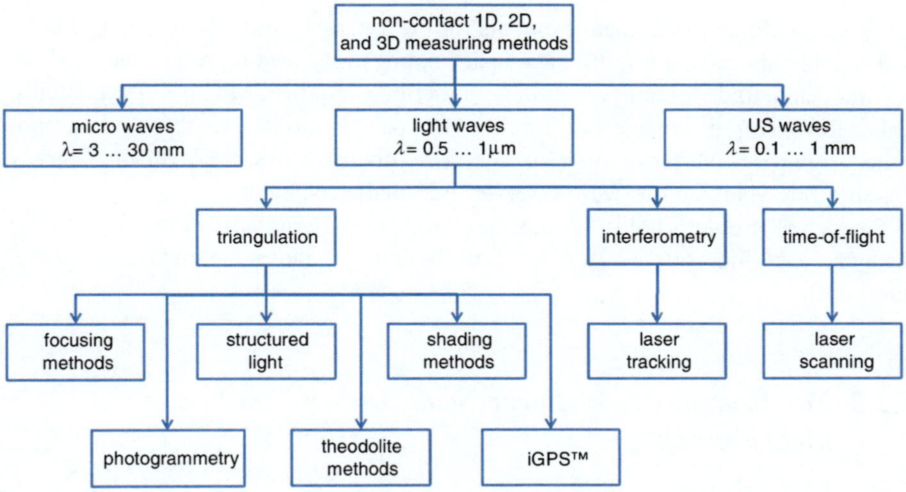

Fig. 1.2 Taxonomy of non-contact measuring methods (Luhmann et al. 2006). (with permission)

- *Triangulation techniques*: photogrammetry (single, stereo, and multiple imaging), angle measuring systems (theodolites), iGPS™, structured light (light section procedures, fringe projection, phase measurement, moiré topography), focusing methods, shadow methods, etc.
- *Interferometry*: optical coherent time-of-flight measurement, holography, laser interferometry, speckle interferometry, coherent radar.
- *Time-of-flight based techniques*: distance measurement by optical modulation methods, pulse modulation, etc.

A taxonomy of these methods is reported in Fig. 1.2.

1.2.1 The Pressing Need for LSM Solutions

Up until the last decades of the past century, engineering metrology was principally restricted to objects of medium to small dimension. Until that time, the term "precision engineering" mainly referred to objects ranging from a few micrometers up to, at the most, some meters. Recent developments in the different fields of engineering have meant a redefinition of this concept. This is mostly due to the necessity for measuring very large structures with high levels of accuracy.

Traditionally, shipbuilding was the field most involved in the measurement of large elements, both in assembly and control of components. In the past, the problem was solved by using "traditional instruments", such as tapes and theodolites. This approach was quite laborious and time-consuming, but it assured a level of accuracy fit for the purpose.

The use of traditional equipment started to become inadequate with the evolution of the aviation industry, which was requiring more and more accurate

Large-Scale dimensional measurements for the assembly and construction of large and reliable aircrafts. Currently, aerospace applications best represent the need for accuracy and, under certain conditions, sometimes require CMM-proof accuracies. Take, for example, capsule construction, or space stations assembly and components alignment. All these large-sized objects involve the same level of accuracy which a few years before were reserved for smaller elements.

The same problem holds for the measurement of large structures, such as big bridges, metallic trusses for special buildings, radio antennas, or radio telescopes.

1.2.2 The Complexity of Large-Scale Dimensional Measurements

When dealing with Large-Scale dimensional measurements the problem arises of how to interface the measured object with the measurement system. It is sometimes impossible to transport a large-sized object to the measuring device. Hence, the traditional CMMs can hardly be employed. It becomes more practical to move the measuring instrument to the measured object. This is the reason why many measuring systems for Large-Scale Dimensional Metrology applications consist in portable or distributed systems, so system itself can be moved and installed all around the measured object.

Furthermore, above a certain size, the structural behaviour of a traditional measuring machine would be so instable as to have a significant influence on the measurement results. Some examples of these conditioning factors are the bending stresses exerted by the weight of the machine itself on its elements, or the inertial effects due to the high mass of machine components.

On the other hand, large-size measurements are heavily influenced by boundary conditions. In many cases, the extension of the measured object is so large that the external influencing factors (humidity, temperature, air pressure and density, atmosphere components, etc.) present significant variations all along the measured shape. These aspects must be under close control during the measurement procedure. Besides, the measured object may warp under its weight and present geometrical deformations and/or structural variations along its shape.

The optical and acoustic approaches for Large-Scale Dimensional Metrology applications, which have completely supplanted the traditional ones, are extremely sensitive to the variations of uniformity in the propagation medium. Hence, when using this systems, appropriate corrections can be made in order to eliminate the induced systematic errors. Many interesting contributions of this kind have been presented in literature (Estler et al. 2002; Maisano et al. 2009).

Another crucial task is to make measurements in the so-called "shadowed zones", i.e. those parts of the measured object where the employment of the measuring system is not easy and immediate. This is the typical case of complex surfaces, characterized by many geometrical singularities and undercuts.

The problem is usually solved by using distributed systems with an opportune positioning of sensors or by the employment of special probes, opportunely shaped in order to be introduced into very narrow holes or crevices. In many cases the whole measuring system must be designed or adapted in order to satisfy shape and dimension constraints.

In most cases the measurement is performed under static conditions: only stationary objects are measured. This means a large frequency of acquisition is not required and the operator can perform the measurement with no need to change the sensor position and/or orientation during operations. Furthermore, data processing software does not need to be equipped with algorithms for cinematic analysis.

The situation is different when Large-Scale measuring systems have to be employed for tracking moving objects. In this case the whole measuring system must be able to acquire large quantities of position data with a very high frequency (at least consistent with the speed of the measured object). Furthermore, this is not the only problem connected to the tracking of moving object. Other topics include coverage of the measuring volume, distinguishing between different moving objects, identification of the moving direction and object attitudes.

A particular kind of cinematic measurement is where the measured object is stationary, but the probe(s) or the measuring device(s) is (are) rapidly moved all around the measuring space in order to obtain the greatest number of acquisitions as possible in a short time.

The topic of cinematic measurements presents a set of specific characteristics, that require a separate description. This subject is not treated in this book.

1.3 Overview and Taxonomy of LSM Systems

Maisano et al. proposed a classification of the major Large-Scale Dimensional Metrology instruments according to the following definitions (Maisano et al. 2009):

- *Centralized systems* a centralized system is essentially a stand-alone unit which can work independently to provide the measurement of a spatial coordinate on the object surface, e.g. a laser tracker. In some cases, a number of centralized systems can be used simultaneously with the aim of improving measurement accuracy.
- *Distributed systems* a distributed system consists in a series of measuring stations that work cooperatively to collect information for determining coordinates of a point on the object's geometry. In general, the individual stations associated with a distributed system cannot measure coordinates separately.
- *Contact systems* a contact measuring system is a metrological system which can provide the coordinate of the object to be measured simply by touching it with a probe. The probe of the metrological system can be moved either manually or by

Table 1.1 Classification of major large-scale measuring instruments (Maisano et al. 2009) With permission

	Centralised	Distributed
Contact	CMM, laser tracker, total station	iGPS^{TM}, HiBall^{TM}, Contact systems with multilateration technique
Non contact	Theodolite, laser radar, tacheometer, optical probe CMM, camera based triangulation	Photogrammetry, non-contact systems with multilateration technique

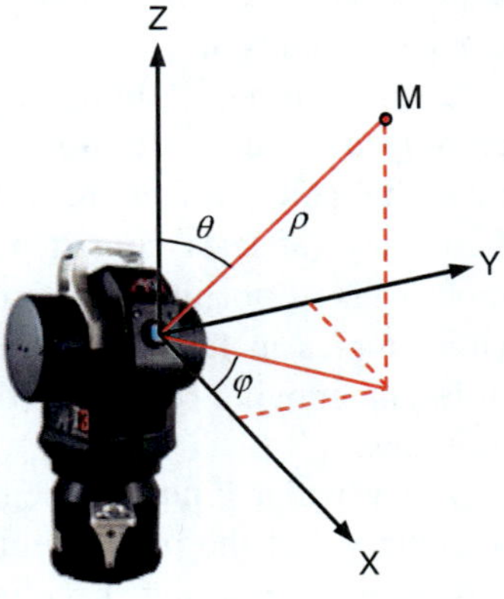

Fig. 1.3 Working principle of laser trackers, laser radars and total stations. The position of a point is defined by a range distance (ρ) and two angular measurements (elevation angle θ and azimuth angle φ) (adapted from Cuypers et al. 2009). (with permission)

mechanical arms or can be attached to the object as a target to be followed by the system.

- *Non contact systems*: these systems can evaluate dimensional features of the object to be measured without the need for a probe to touch the object. They are mainly based on optical technologies.

Table 1.1 presents the major Large-Scale Dimensional Metrology instruments classified according to the proposed taxonomy.

Large-Scale measuring systems presented in Table 1.1 can be further classified depending on their working principle (Cuypers et al. 2009):

- *Measuring systems using two angles and one length*: Most of the Large-Scale measuring systems rely on the determination of one length and two angles (see Fig. 1.3). In these systems the initial coordinates of a point are evaluated in a spherical coordinate system (ρ, θ, φ).
  These systems are also called Spherical Coordinate Measurement Systems. For each system, the angles are measured by means of angular encoders, whilst the range measurement can be performed using either an interferometer like laser trackers or an ADM (Absolute Distance Measure) like laser radars and total stations, or a combination of both technologies like ADM enabled laser trackers. The spherical coordinates are easily transformed in Cartesian coordinates by a central processing unit that is able to derive the object features from the measured points in the 3D space.

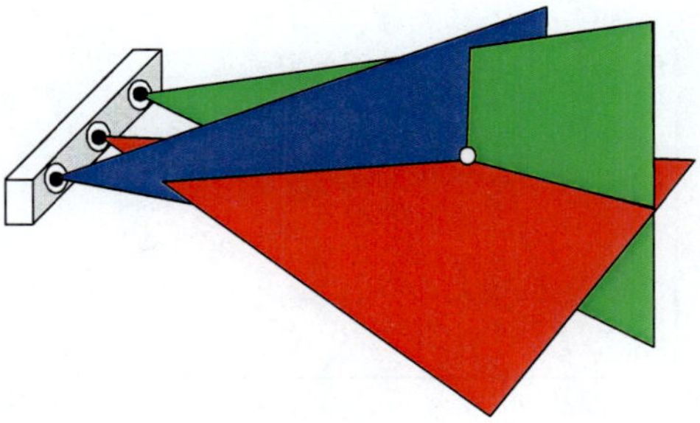

Fig. 1.4 The working principle of camera-based triangulation with 3 linear CCD cameras, resulting in 1 horizontal and 2 vertical angle measurements. The position of the target is determined by planes intersection (adapted from Cuypers et al. 2009). (with permission)

- *Measuring systems using multiple angles (triangulation)*: Instead of using two angles and a distance measurement, it is possible to evaluate the position of a point in a three dimensional space using only angular information from two or more reference points. This working principle relies on the very well known triangulation algorithms. Triangulation uses the known locations of two or more reference points, and the relative angles between the point to be localized and each reference point. In this case the unknown position of the point can be found by solving a linear system (Doğançay 2005).
  A camera-based triangulation system applies this principle. Three linear CCD (Charge Coupled Device) cameras look at targets in space and each camera determines one angle and thereby one plane that contains the target. By using three cameras, the resulting position of the target can be calculated based on the three measured angles. This can be visualized as an intersection of three planes (see Fig. 1.4).
  The working principle of the iGPS™ by Metris is also based on multiple angle measurements. If the horizontal and vertical angles from two or more transmitters are known, the system univocally determines the position of a positioning target. In order to obtain accurate angle measurements the iGPS™ uses rotating laser beams (ARC Second 2010). This system will be described in more detail in Chap. 2.
  Photogrammetry is a Large-Scale measurement technique based on angle measurements. The principle is similar to that of camera-based triangulation (Mikhail et al. 2001) (see Fig. 1.5).
- *Measuring systems using multiple lengths (trilateration)*: Trilateration uses the known locations of three or more reference points, and the measured distance between the point to be localized and each reference point (see Fig. 1.6). The unknown coordinates can be found by solving a non linear optimization problem. This approach is very similar to GPS (Global Positioning System) localisation principle (Hofmann-Wellenhof et al. 2001). Multilateration principles are used by the measurement systems based on laser interferometers as well as by those based on TOF (Time-of-flight) or TDoA (Time Difference of Arrival) (Cuypers et al. 2009).

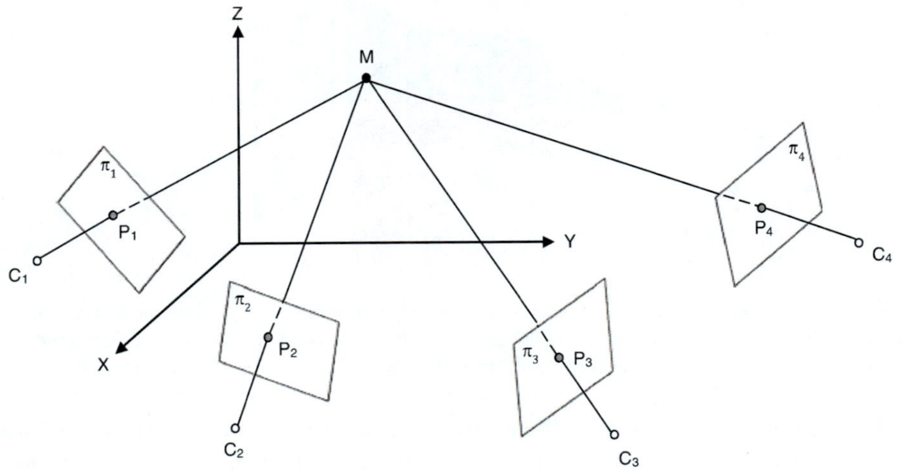

Fig. 1.5 Working principle of camera-based photogrammetry. With knowledge of the 3D spatial position of each camera C_i (with $i \geq 2$), its orientation and technical characteristics (e.g., focal length, lens distortion), the 3D position of a generic point M can be reconstructed according to its 2D positions P_i in the camera view planes (π_i) (adapted from Mikhail et al. 2001). (with permission)

Fig. 1.6 The working principle of trilateration in two dimensions. Knowing the distances from three different reference points (V_1, V_2, V_3) with known coordinates, it is possible to localize the position of an unknown point on the intersection of three circumferences

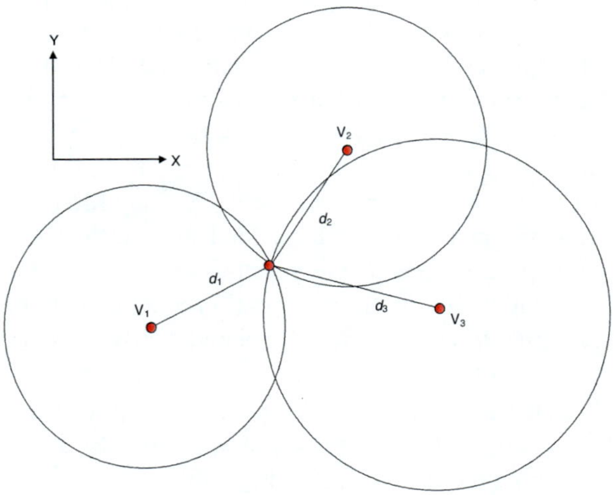

1.4 Distributed System Philosophy

As discussed in the previous sections, Large-Scale Dimensional Metrology could be said to apply to any dimensional measurement where the metrology instrument has to be brought to the object rather than vice versa.

As well as classical centralized metrology systems, in which a stand-alone unit can provide geometrical features of an object to be measured, the latest advances in Large-Scale Dimensional Metrology offer distributed solutions in which a network of metrological stations is implemented to share the measurement task: a portable positioning target is localized relying on all the measurements obtained by the metrological stations. The new metrology approach is the result of a continuous effort toward scalable technologies able to cover flexible working spaces. Classical centralized approaches to Large-Scale Dimensional Metrology, such as laser trackers, theodolites or gantry CMMs can often be impractical when dealing with complex working volumes or dimensions of hundreds of meters. Distributed metrology systems, on the contrary, can easily shape the working volume by opportunely adding one or more metrological station. Although flexibility and scalability are probably the major advantages of distributed metrology systems, their metrological performances are still hardly comparable to those of centralized systems.

Distributed systems for Large-Scale Dimensional Metrology applications present a set of peculiarities in comparison to other systems that is worth discussing in detail.

From the strictly metrological point of view, each distributed system can be seen as a single measurement instrument. Hence, it must be calibrated, its metrological performance (e.g. stability, repeatability, reproducibility, and accuracy) must be tested, and its reliability must be verified.

On the other hand, the overall behaviour of distributed systems is strictly connected to the specific performance of each single component, which must be tested both individually and globally, that is when it interacts and/or cooperates with the other elements of the whole system.

Other topics, such as the measuring volume coverage and the components cooperation, are typical aspects of these systems, and must be specifically investigated either during the design or during their implementation.

1.4.1 Coverage of Measuring Volume

One of the most attractive properties of distributed systems is the possibility of allocating the distributed sensor devices in order to cover the whole measuring volume. This is a great advantage especially in comparison to other non-distributed systems (see, for example, laser trackers or theodolites), which require to move the measuring device from its actual position when the object features to be measured are positioned in a shadowed zone or are not directly accessible. In this case, the main problem is to keep the same coordinate reference system for measures acquired in different positions of the measuring device. On the contrary, with distributed systems, once the sensors are placed and the system is calibrated, in principle, no change of network configuration is needed. This guarantees the same common reference to all the measured points. In some special situations, if some network sensors need to be added, removed or reallocated, the realignment to the original reference system is simple and easily automatable.

That said, the challenging aspect of sensor allocation is how to find an optimal configuration, which assures the total coverage of the measuring regions, with the required level of accuracy, and the appropriate number of sensors.

Different studies presented in literature have produced some semi-automatic procedures and algorithms for optimal sensors positioning (Galetto and Pralio 2010). In Chap. 4 an analysis of the state-of-the-art of the research indifferent application fields and a case study applied to a distributed metrological system are presented.

The main consideration regarding all the approaches reported in literature is that the proposed algorithms may represent an effective support for the operator in network device arrangement, but their use cannot ignore an empirical validation of the suggested configuration.

1.4.2 On-Line Self-Diagnostics

The concept of "on-line self-diagnostics" is closely related to the notion of on-line metrological performance verification in traditional automated systems. This approach is complementary to that of uncertainty evaluation. The on-line verification acts as a guarantee for the preservation of the measurement system characteristics (including accuracy, reproducibility and uncertainty).

Techniques for "on-line metrological performance verification" individuate anomalous behaviour in measuring systems, and automatically reconfigure the control system to assure the preservation of metrological characteristics.

System degradation produces non-reliable measurements. The indication of possible faults can be followed by more accurate tests or by a complete calibration (Franceschini et al. 2009a).

The problem of "self-diagnostics" in automated systems is not a recent matter, and traditionally many strategies have been proposed. In many important areas, such as for example the aeronautical and nuclear sectors, the most widely used techniques are based on redundancy, i.e. the replication of instruments and control equipments. This approach is usually very expensive. In general, two main types of redundancy are found (Franceschini and Galetto 2007):

- "Physical" redundancy, which consists of replicating instruments and control equipment (for example, by integrating the touching probe with an optical device or by making the probe redundant).
- "Analytical" or "model-based" redundancy, which substitutes the replication of a physical instruments by the use of appropriate mathematical models, which may derive from physical laws applied to experimental data or from self-learning methods (for example, neural networks).

Sometimes the two approaches are implemented together. In this case the analytical redundancy method is enhanced with the use of an external "witness-part" (Franceschini et al. 2002).

The topic of on-line diagnostics for Large-Scale distributed systems is treated in Chap. 6, and some application cases are described and analyzed in depth.

1.4.3 Metrological Performance Indicators

The metrological performance of a measuring system is usually tested through the use of specific parameters, generally prescribed by international standards.

Since distributed systems for Large-Scale Dimensional Metrology applications have been introduced in relatively recent years, a consolidated reference set of standards has still to be defined. For this reason, usually, researchers and manufacturers refer to general metrology standards, or current standards for CMM or optical systems (Peggs et al. 2009).

In general, according to the International Vocabulary of Metrology (VIM) (JCGM 200:2008 2008) and the Guide to the Expression of Uncertainty in Measurement (GUM) (JCGM 100:2008 2008), the basic performance indicators for a measuring system are:

- *Accuracy*: closeness of agreement between a measured quantity value and a true quantity value of a measurand.
- *Uncertainty*: non-negative parameter characterizing the dispersion of the quantity values being attributed to a measurand, based on the information used.
- *Repeatability*: closeness of agreement between indications or measured quantity values obtained by replicate measurements on the same or similar objects over a short period of time under specified conditions of measurement, out of a set that includes the same measurement procedure, operators, measuring system, operating conditions and location.
- *Reproducibility*: closeness of agreement between indications or measured quantity values obtained by replicate measurements on the same or similar objects under specific conditions of measurement, out of a set that includes different locations, operators and measuring systems.
- *Precision*: closeness of agreement between indications or measured quantity values obtained by replicate measurements on the same or similar objects under specified conditions (the term *precision* is also used to indicate *repeatability* and *reproducibility*).
- *Stability*: property of a measuring instrument, whereby its metrological properties remain constant over time.
- *Resolution*: smallest change in a quantity being measured that causes a perceptible change in the corresponding indication.

Every measuring system must be characterized by evaluating these parameters. If, after the test, the results are in agreement with the required values, the system is utilized as it is, otherwise some adjustments or corrections have to be made.

Traditional approaches for CMMs and optical instruments prescribe performance evaluation and uncertainty estimation.

In general, performance evaluation refers to the assessment of accuracy, repeatability and reproducibility of the measurement performed in a well-defined point (or portion of volume) in the whole measurement domain of the system. To this end, current standards prescribe the use of specific calibrated artefacts and operational procedures, implemented in order to identify all the possible sources of error emerging during the measurement (Peggs et al. 2009).

On the contrary, uncertainty estimation refers to the evaluation of the uncertainty associated with measured data and related measuring procedure. In Large-Scale Dimensional Metrology, uncertainty evaluation involves developing a model of the random and systematic effects associated with the measuring instrument(s) and analyzing how they affect the uncertainties associated with the estimated point coordinates. The Guide to the Expression of Uncertainty in Measurement (GUM) (JCGM 100:2008 2008) prescribes an approach based on the multivariate law of propagation of uncertainty.

A specific discussion and application of the multivariate law of propagation of uncertainty is presented in Chap. 8, while several applications of performance evaluation and systematic error correction are reported in Chap. 3 and Chap. 7, respectively.

1.4.4 Wireless Sensor Networks (WSNs) and Distributed Systems

WSNs are typically composed by small and lightweight devices that can be easily deployed and arranged in a working environment. Furthermore, each device is generally provided with both communication and computation capabilities given by the embedded electronic components. These features certainly increase the appeal of WSNs and make them suitable for the design of a fully distributed system for dimensional measurements.

A distributed network-based layout provides a system with profitable scalability features, peculiar to measurement systems based on spatially distributed sensing units (Nikon Metrology 2010; 3rdTech 2010). As a matter of fact, modular architecture makes such systems suitable for Large-Scale Dimensional Metrology applications, overcoming limitations of existing digital photogrammetry-based systems (Axios 3D 2010; Metronor Corporate 2010; Northern Digital 2010). Real-time coordinate acquisition of different targets, possibly located in different regions of the working volume, is then possible by spreading the sensor devices around, provided that the acquisition task is synchronized and a common reference system is given. These capabilities make distributed systems feasible solutions for tracking mobile objects, even if they are characterized by fast dynamics. This property is particularly interesting in an attempt to automate the contact measurement procedure. Most commercially available instruments provide a hand-held probe for touching the reference measurement points (Automated Precision 2010; Leica Geodetic Service 2010; Axios 3D 2010; Nikon Metrology 2010), thus involving direct interaction between the sensor equipment and the human operator as well as a strong dependence on his/her skills. An alternative approach relies on autonomous unmanned platforms for carrying the sensor equipment and moving the contact probe around the working volume (Franceschini et al. 2010a). This new perspective, scaling down the human role to a mission management task, clearly shows the need for a system able to perform measurements for control as well as for metrological issues.

As described in Sect. 1.2, currently available dimensional metrology systems rely either on distance or angle measurements. Thus the possible use of a WSN-based system for dimensional metrology applications is certainly bound by its capabilities of performing such kinds of measurements. Nowadays there are many approaches to this field, relying on different technologies and sensors. Angular measurements can be achieved, for example, using accelerometers, magnetometers, gyroscopes, CCD sensors, photodiodes or simply measuring the difference in the received phase of a Radio Frequency (RF) signal at each element of an antenna array (Kwakkernaat et al. 2008). On the other hand, distance measurements can be obtained, for instance, by evaluating the time-of-flight of particular signals (such as US signals), the time difference of arrival of different signals or the received strength of a RF communication signal (Franceschini et al. 2009c).

Whatever the system components and the localisation algorithms are, a WSN-based metrology system represents a further step towards hardware and software automation in dimensional measurement applications. Due to its capabilities of sharing the metrology task, each network device works cooperatively with the aim of determining the geometrical features of an object. In this way, the measurement results as the synthesis of the information gathered locally and shared by each network node. Communication links among network nodes also provide the possibility of reconfiguring their orientation during the task according to measurement conditions and procedures, aiming at optimizing the overall system performance.

1.4.5 Localisation Algorithms and System Calibration

When dealing with distributed measuring systems, the choice of an appropriate method for localizing network devices and a technique for system calibration is crucial. As mentioned in Sect. 1.4.4, it could be helpful to think of a distributed system as a wireless network of sensors (Franceschini et al. 2009a). In general, a wireless network typically consists of a large number of nodes (equipped with sensing devices and transceivers) with a dense distribution, equipped with transceivers. Each device can communicate with other devices within its communication range. A wireless network is typically modelled as a graph, where each node represents a physical device. Two nodes are connected by an edge, if and only if they can directly communicate.

Dramatic advances in integrated circuits and RF technologies have made possible the use of large WSNs for many applications. In particular, the utilization of WSNs in metrology is attracting more and more attention. Since sensor devices do not need cables and may be easily deployed or moved, they can be utilized for a variety of industrial metrology applications.

In order to perform metrological measurements, fixed network nodes should be aware of their respective locations. To achieve this, especially for Large-Scale sensor networks, many self-localisation methods have been studied and implemented recently. Generally, nodes automatically cooperate, estimating local distances from their neighbours, converging in a consistent coordinate assignment. Nodes work together in a peer-to-peer way to build a map of the sensor network.

Received-Signal-Strength (RSS) and *Time-of-Arrival* (ToA) are two common approaches for estimating the distance between nodes within their mutual transmission range (Wu et al. 2008). RSS measures the power of the signal at the receiver and calculates the distance according to the propagation loss model. ToA measures the propagation time (Δt) of the received signal (typically a RF signal for large distances or US for small distances) and determines the distance by multiplying it by its own speed. In general, RSS is easier to implement, while ToA may achieve higher accuracy (Patwari et al. 2005).

Angle of Arrival (AoA) is another approach for WSNs localisation. Usually, sensor nodes receive the signals from at least three neighbours—in particular, collecting the angle information—and determine their coordinates by triangulation according to the angle bearings of incoming signals (Nasipuri and Li 2002; Niculescu and Nath 2001).

Generally, localisation algorithms are designed to be applied to a typical sensor network, consisting of a large number of nodes with a dense distribution. As a consequence, many of them do not fit for small networks, with few distributed nodes. In such cases, nodes can be manually located.

Localisation algorithms can be classified according to four different classification criteria:

1. The first classification is based on the presence (or absence) of nodes with pre-configured coordinates.

    - *Anchor-based algorithms* The localisation system is implemented starting from a set of reference nodes ("landmarks", "anchor-nodes") with known coordinates.
    - *Anchor-free algorithms* These use local distance measurements among nodes to determine their respective coordinates. They do not assume the availability of nodes with pre-configured coordinates.

2. The second classification is based on the way node localisations "propagate" in the network.

    - *Incremental algorithms* These algorithms usually start with a set of three or more reference nodes with known coordinates. Other nodes in the network can contact the reference nodes and determine their own coordinates. When a node at an unknown position obtains an acceptable position estimate, it can serve as a new reference point. This process can be applied incrementally until all nodes in the network have obtained their coordinates.

- *Concurrent algorithms* In this approach, many pairs of sensors communicate and share measurements, in order to achieve localisation for all sensors. Rather than solving each sensor position one at time, all sensor positions are simultaneously estimated ("cooperative systems").

3. The third classification subdivides localisation approaches into two broad classes, based on the "granularity" of information acquired by the sensors during communication.

   - *Fine-grained algorithms* Algorithms that use accurate information—such as the distance from a reference point based on RSS or ToA measurements—fall into the category of fine-grained localisation methods. Typically, they use technologies, such as IR, US, or RF signals.
   - *Coarse-grained algorithms* Algorithms that utilize less accurate information, such as proximity to a given reference point, are categorized as coarse-grained localisation methods.
     Coarse-grained algorithms estimate inter-node distances using rough techniques like hop-count. In a wireless network, the number of hops is the number of edges traversed by a signal, along the shortest path between the source node and the destination node. Hop-count may be used to determine a rough evaluation of inter-node distances (Priyantha et al. 2005).
     As expected, fine-grained algorithms are more accurate than coarse-grained. In the absence of measurement errors, fine-grained algorithms provide an exact positioning of network nodes.

4. The fourth classification is based on computational distribution.

   - *Centralized algorithms* Computational load is performed by a single centralized node or network device. All nodes broadcast information to a single computer to solve the localisation problem.
   - *Distributed algorithms* Computational load is equally distributed among network nodes. Each node receives localisation information from neighbouring nodes, performs computation, and transmits the obtained results back to them.

According to the principles of *Received-Signal-Strength* (RSS), *Time-of-Arrival* (ToA), or *Angle of Arrival* (AoA), many specific algorithms for sensor network calibration and measured point coordinates evaluation may be implemented. A description of some specific algorithms for point coordinates assessment is reported in Chap. 2 for iGPS^{TM}, and in Chap. 3 for MScMS-I and MScMS-II. System calibration procedures are reported and discussed in Chap. 5.

1.5 Advantages and Weaknesses of Distributed Systems

The appeal of distributed systems derives from many features that make them different from conventional centralized systems:

- *Flexibility* As they consist of multiple remote sensors, distributed systems can easily be arranged in the working volume according to user needs, the geometry of the environment, and the measurement task. System flexibility can be further enhanced by implementing pre-processing software tools to find possible configuration layouts, in order to optimize the metrological performance and/or the measurement volume (Galetto and Pralio 2010). The possibility of adding or removing sensing units according to user needs makes these systems extremely flexible as to their implementation for industrial applications.
- *Redundancy* In typical working conditions, distributed systems are often able to refer to more distributed components than strictly necessary. Depending on the localisation technique adopted, information redundancy enhances system accuracy and gives the system the possibility to implement real time verification strategies.
- *Reliability* Reliability is the ability of a system to perform and maintain its functions in routine circumstances, as well as in hostile or unexpected circumstances. If one or more remote devices is not working properly, distributed metrology systems, generally characterized by hardware redundancy, can actually use the "healthy" nodes to compensate for the malfunctioning of a part of the network.
- *Scalability* The main strong point of a distributed system is the capability to adapt easily to large dimensions and unusual shapes. The real working volume of a distributed metrology system is related to the network layout. Changing density and/or position of the remote sensing units, the user can size and shape the working volume, within the network design phase as well as during the experimental campaign.
- *Concurrent measurement capability* Distributed metrology systems generally allow the use of different measurement tools (multiple targets and/or portable probes) at the same time. Once the system infrastructure has been set up, an unlimited number of tools can actually operate within the work space, without any additional cost per user.
- *Sensor fusion* The metrological system can be integrated with other spatially distributed sensors (in order to monitor temperature, humidity, vibrations, light intensity, etc.). Sensor data fusion makes it possible to perform an environmental mapping of the working volume and to monitor the operating conditions of the dimensional measuring devices.

On the other hand, unlike centralized systems, the distributed nature of these systems requires the coordination and management of multiple stations. At the moment, the main disadvantages of these systems are:

- *Set-up* In order to work properly, every distributed system needs to know several parameters of the local hardware. Some of these parameters may change because

of environmental factors such as vibrations, thermal change or other accidental reasons. In order to achieve optimum accuracy, each distributed system generally needs a careful set-up phase. During this phase, which can be automated to some extent, the system calculates information like sensor positions and orientations, local temperatures, humidity, pressure and so on. This information is useful during the measurement. Errors during the set-up phase adversely affect the accuracy of the measurements (Maisano and Mastrogiacomo 2010).
- *Expertise* Distributed metrology systems are typically less user friendly than centralized systems. They generally need a more experienced and careful user, especially during the set-up process. Since they consist of multiple stations, particular attention has to be paid to coordinating the data acquisition from different sensing devices (e.g., sensor device synchronization).
- *Standards* While these new systems are attractive to potential end-users, standards, best practice methods and independent performance verification techniques are at present very limited (Peggs et al. 2009).
- *Accuracy* The performance of distributed metrology systems is strongly related to several factors that can affect the accuracy of the system adversely, such as the number of network devices, the set-up parameters, and the relative position in the working volume of the points to be measured.

1.6 Some Examples of Distributed Systems

As introduced in the previous sections, distributed systems consist of a series of measuring stations that work cooperatively with the aim of determining object geometrical features. In general, the individual stations associated with a distributed system cannot measure coordinates as stand-alone units.

Among all the Large-Scale Dimensional Metrology instruments classified above, the distributed instruments are the most recent and for this reason they are the ones which show the greatest potential and are still subject of research. Although photogrammetry is a mature and well known distributed technique, its real potential is still to be discovered due to the recent advances in computer performance within the last decade (Mikhail et al. 2001).

All distributed systems have a similar common architecture (see Fig. 1.7):

- A network of multiple sensors distributed around the working volume.
- A wireless portable probe equipped with sensors able to detect the signals received from the distributed sensors or a set of target sensors to be localized. The portable probe can be moved either manually by a human operator or automatically by a robot.
- A processing unit able to process the data streaming sent by the portable probe or the network elements.

Depending on the measurement capability of the measurement stations, distributed systems can measure using multiple angles or lengths. In the following, by

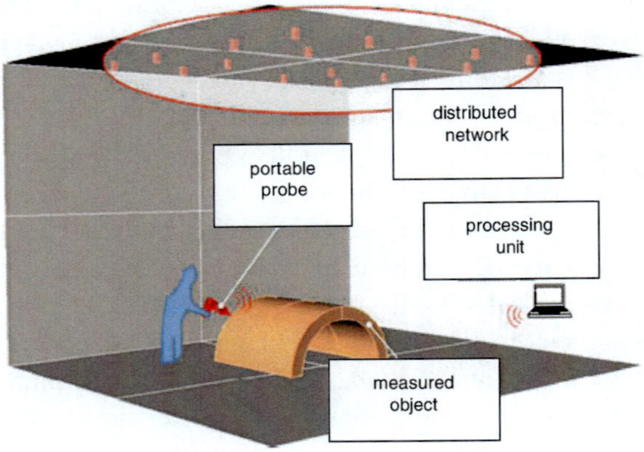

Fig. 1.7 Classical layout of a distributed metrology system (adapted from Galetto et al. 2011). (with permission)

way of example, two of the more relevant distributed metrology solutions are briefly described: the Hiball™ and a Photogrammetric System. They represent two typical instances of implementation of optical technology in order to obtain dimensional measurement with the triangulation technique.

Among the other distributed systems, iGPS™ will be described in detail in Chap. 2, while the two innovative systems MScMS-I and MScMS-II will be presented in Chap. 3.

1.6.1 HiBall™

The HiBall™ Tracker is a new approach to wide-volume tracking and measuring, delivering accuracy with low latency, high update rate, and scalability to cover a large region. It is composed of two key integrated components; the HiBall™ Optical Sensor mounted on a portable probe and the HiBall™ Ceiling Beacon Arrays (the network components). The HiBall™ Optical Sensor is composed of 6 lenses and photodiodes arranged so that each photodiode can 'view' IR LEDs (Light Emitting Diodes) of the Beacon Arrays mounted on the ceiling, through several of the six lenses (see Fig. 1.8), (Welch et al. 2001).

The portable probe estimates the angles from the HiBall™ Beacon Arrays seen by the photodiodes. The position of the probe is found by triangulation given that the localisations of the Beacon Arrays are known. Although the system needs the localisation of the HiBall™ Beacon Arrays, no special adjustments are required for the ceiling structure—the system precision is unaffected by typical variations in ceiling height or in strip placement. The self-calibration feature of HiBall™ system then rapidly determines the localisation of the individual ceiling strips creating a "ceiling map" that can be saved and/or continuously updated while tracking (Welch et al. 2001).

This system works at a frequency of about 2,000 Hz, suitable also for accurate tracking even with rapid movements. The localisation resolution is higher than 0.2 mm (Welch et al. 2001) with an angular accuracy of more than 0.01°.

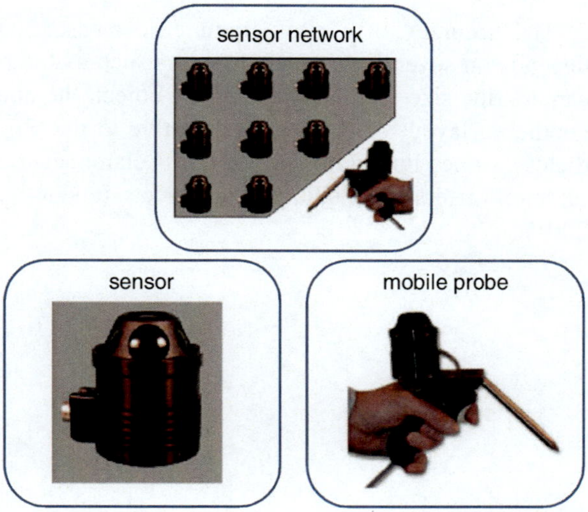

Fig. 1.8 The HiBall™ system. The network components are embedded in a series of ceiling mounted strips while the probe is equipped with photodiode sensors (adapted from 3rtTech 2010). (with permission)

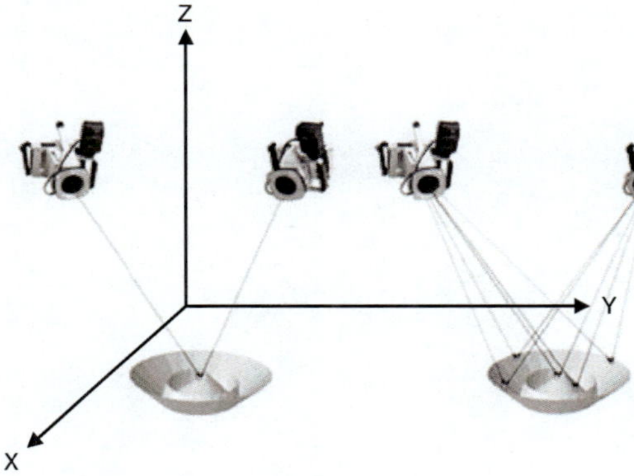

Fig. 1.9 Single and multiple point triangulation using photogrammetry

1.6.2 Photogrammetry

Photogrammetry, as its name implies, is a 3-dimensional coordinate measuring technique that uses photographs as the fundamental medium for metrology (or measurement) (Cuypers et al. 2009).

The fundamental principle used by photogrammetry is triangulation. By taking photographs or video images from at least two different locations, so-called "lines of sight" can be developed from each camera to points on the object. These lines of sight (sometimes called rays owing to their optical nature) are mathematically intersected to produce the 3-dimensional coordinates of the points of interest (see Fig. 1.9).

In order to triangulate a set of points the camera position and aiming angles must be known for all the pictures in the set. Some commercial photogrammetry systems do this operation automatically together with the localisation of the targets on the object to be measured.

The accuracy of a photogrammetric measurement can vary significantly. It depends on several interrelated factors such as the resolution (and quality) of the camera, the size and the shape of the object, the number of images taken and the geometric layout of the cameras relative to the object and to each other. Nevertheless, some photogrammetry systems claim accuracies of the order of some tens of micrometers on medium size objects (about 3 m of length) (Mikhail et al. 2001).

Chapter 2
Indoor GPS (iGPS^{TM})

2.1 System Architecture

Indoor GPS (iGPS^{TM}) is a modular, large-volume tracking system enabling factory-wide localisation of multiple objects with metrological accuracy, applicable in manufacturing and assembly. The system components of iGPS^{TM} are a network of transmitters, a control centre and a number of wireless sensors (Kang and Tesar 2004). The distributed nature of the system makes handling easier and provides scalability for the coverage of the measuring area. For this reason, iGPS^{TM} is more suitable for particular types of measurement, which cannot be carried out by conventional instruments, like Coordinate Measuring Machines (CMMs). For instance, some large-size objects can not be transferred to the measurement systems due to their dimensions or other logistic constraints. For the system operator, iGPS^{TM} can potentially be considered as a faster and easier solution compared to conventional CMMs, theodolites or laser trackers.

Transmitters (see Fig. 2.1) use laser and infrared light to determine the relative angles from the transmitters to the sensors. Sensors, used for measuring the workpiece, have photodiodes inside their modules that can sense the transmitted laser and infrared light signals. Based on the known location of the transmitters, which is normally obtained in an initial network calibration phase, the position of the sensors can be subsequently calculated. The signal is transferred through a wireless network connection providing mobility to the operator. Similar to a classical satellite-based GPS, a one-way signal path is created from transmitters to each sensor. This approach allows an unlimited number of sensors to continuously and independently calculate positional data. A short comparison between satellite-based GPS and iGPS^{TM} is presented in Table 2.1. For a detailed description of the satellite-based GPS, we refer the reader to the literature (Hofmann-Wellenhof et al. 2001).

Going back to the description of iGPS^{TM}, measurements are taken by touching the required points on the object's surface with a probe that is equipped with a pair of sensors. Points are defined on a Cartesian coordinate system; the coordinates are

DOI: 10.1007/978-0-85729-543-9_2,

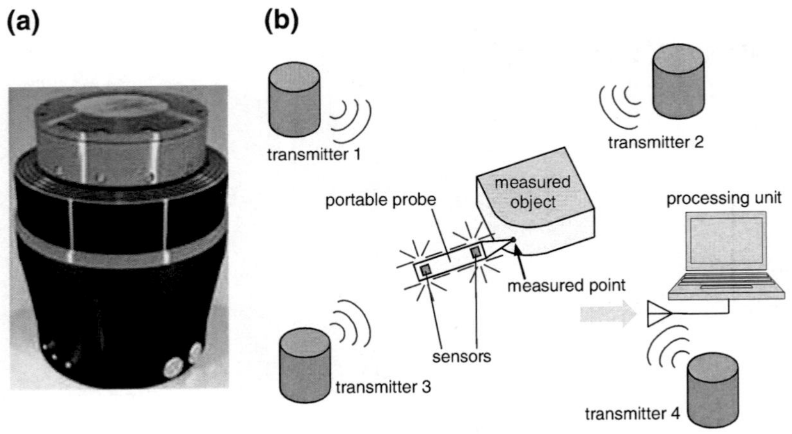

Fig. 2.1 **a** iGPS™ transmitter (Nikon Metrology 2010). **b** Representation scheme of an iGPS™ network of transmitters and its portable probe (Maisano et al. 2008). (with permission)

Table 2.1 Short comparison between the classical satellite-based GPS and the indoor GPS

	Satellite-based GPS	Indoor GPS
Network infrastructure	Over 30 satellites in orbit above the Earth's surface	Transmitters installed around the object to be measured
Signals sent out from transmitters to sensor(s)	Radio-frequency signals	Infrared laser-pulses
Data used to localize sensor(s)	Instant position of satellites and corresponding distances from satellites to sensor(s)	Position of transmitters and relative angles from transmitters to sensor(s)
Localisation technique	Trilateration	Triangulation

then processed by specific algorithms in order to determine geometric features. These measured features are then used to extract the desired dimensional information such as feature positions and angles between two features (ARC Second 2010).

2.2 Working Principles

Typically, the system components of iGPS™ are two or more transmitters, a control centre and a number of wireless sensors.

Transmitters operate as reference points (with known position) continually generating three signals: two infrared laser fanned beams rotating in the head of the transmitter and an infrared LED strobe (ARC Second 2010; Maisano et al. 2007, 2008, 2009). Sensors are wireless elements, which can be mounted on a

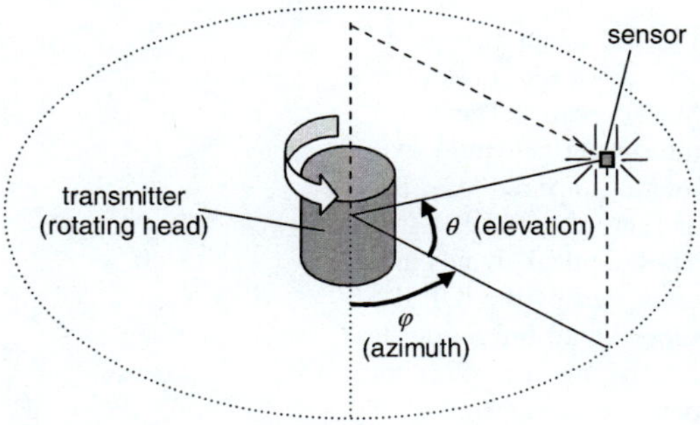

Fig. 2.2 Azimuth (φ) and elevation (θ) angles from a transmitter to a sensor (Maisano et al. 2008). (with permission)

portable hand-held probe or placed on the surface of the object to be measured, in order to receive the transmitters' signals.

iGPS^{TM} is a scalable (or modular) system since the number of transmitters and sensors can be increased according to the measurement environment. Such characteristic, however, does not compromise the network communication or slow down network calibration activities and measurements (ARC Second 2010).

The location of transmitters has to be determined before starting measurements. This phase should be fast and as automated as possible to prevent any conflict with the system's adaptability to different working environments.

During measurements, for each sensor the position (x, y, z) is calculated. Each transmitter presents two measurement values to each sensor: the horizontal (azimuth, φ) and the vertical (elevation, θ) angles (see Fig. 2.2). Sensors can calculate their position whenever they are localised in the line of sight of two or more transmitters. The principle used is triangulation (Niculescu and Nath 2003).

Sensors measure angles from the transmitters as follows. Each transmitter generates two rotating infrared laser beams and an infrared LED strobe. These optical signals are converted into timing pulses through the use of a photo detector. The rotation speed of the spinning head in each transmitter is deliberately set to a different speed in order to differentiate the transmitters. Additionally, the transmitter speed is continuously tracked and used to convert the timing intervals into angles. As shown in Fig. 2.3, the two fanned beams radiated from the rotating head of each transmitter are tilted with respect to the rotation axis (the vertical axis of the transmitter), nominally at $\phi_1 \approx 30°$ and $\phi_2 \approx -30°$. This angular method is used to calculate the elevation angle (θ) by using the following data:

- the angles (ϕ_1 and ϕ_2) of the fanned beams with respect to vertical, as shown in Fig. 2.3;
- the difference in timing (Δt) between the arrival of laser 1 and laser 2 to the sensor;
- the speed of rotation of the transmitter (ω), which is continually tracked.

The measurement of the azimuth angle requires a local reference direction, which is created by firing an omnidirectional LED strobe at a fixed direction in the rotation of the transmitter's head. Referring to the timing diagram at the bottom of Fig. 2.3, the azimuth angle is determined by:

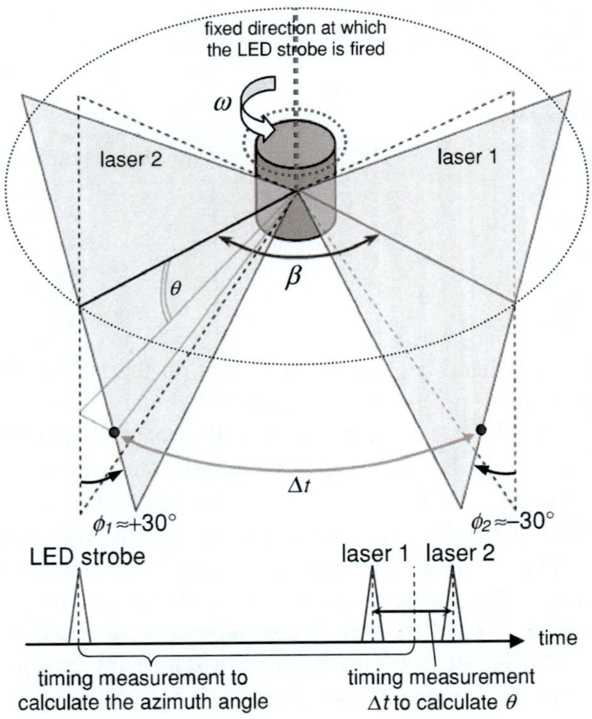

Fig. 2.3 Representation scheme of the transmitter's fanned beams. The transmitter generates two rotating infrared laser beams and an infrared LED strobe. These optical signals are converted by the iGPS™ sensors into timing pulses through the use of a photo detector. In the timing diagram at the *bottom* of Figure, the azimuth angle (θ) is determined (Maisano et al. 2008; Nikon Metrology 2010). (with permission)

- knowing Δt;
- making a timing measurement between the strobe and the laser pulses;
- knowing the speed of rotation of the transmitter.

In addition to the azimuth and elevation angles from the transmitter to the sensor, more information is needed to perform a sensor position calculation, which gives the relative position and orientation of the transmitters.

Transmitters make a network of reference points localized through a calibration process. The relative position and orientation of the transmitters are determined using an advanced algorithm, which is known as bundle adjustment (Hedges et al. 2003; Chen et al. 2003; Triggs et al. 2000). An additional component of network calibration is the "system scale", which is the absolute distance between two known points such as the length of a reference bar. iGPS™ provides a relatively rapid and semiautomated localisation procedure, requiring relatively few manual measurements (Akcan et al. 2006). For more information about the bundle adjustment and the system scaling procedures, see Chap. 5.

Once the network calibration has been completed, the measurements can be performed using a portable handheld measurement probe , known as a V-bar. This probe, equipped with two sensors (Figs. 2.1, 2.4), is held by an operator in order to measure the coordinates of the points touched by the probe tip. To ensure that it is stable and insensitive to thermal expansion, the portable probe is mainly made of composite material.

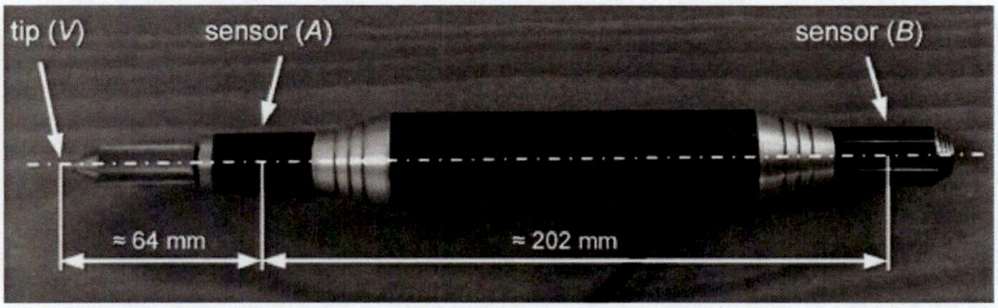

Fig. 2.4 iGPS™ portable hand-held measurement probe (V-bar) (Maisano et al. 2008; Nikon Metrology 2010). (with permission)

In summary, the measurement procedure is made up of three main steps:

- Spatial localization of each sensor (*A* and *B*) is achieved using a triangulation technique. To uniquely determine the relative localization of a point in a 3D space, at least two transmitters are needed (Chen et al. 2003; Akcan et al. 2006). All information needed for the localization is sent to a PC.
- As shown in Fig. 2.4, the probe tip (*V*) lies on the line that connects sensors *A* and *B*. Therefore the localization of the point touched by the probe tip can be calculated using the coordinates of points $A \equiv (x_A, y_A, z_A)$ and $B \equiv (x_B, y_B, z_B)$ and the geometrical features of the probe (distances d_{A-V} and d_{A-B}). The algorithm in use is described in Sect. 3.1.
- Similar to CMMs and laser trackers, it is possible to reconstruct shapes and geometries of objects using suitable software applications. Geometries include cylinders, planes, circumferences, cones, spheres, and any other standard features. This is achieved using a set of measured points from the part surface. Such points are collected using the portable probe, and processed using classical optimization algorithms (Overmars 1997).

2.3 Factors Affecting Measurements

Many factors can affect the quality of a measurement. The most significant include:

- number of transmitters;
- sensor movement during measurement;
- network calibration;
- environmental factors.

2.3.1 Number of Transmitters

The number of transmitters is strictly related to their communication range and the measurement volume. Since the communication range of each transmitter is around 30 m, transmitter density within the measuring volume does not have to be

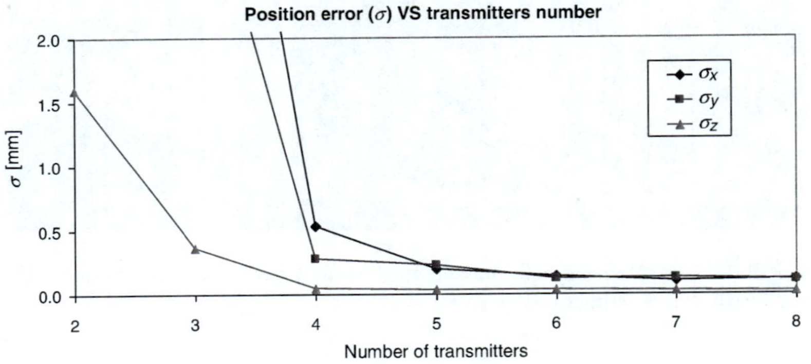

Fig. 2.5 Influence of the number of transmitters on the position error, analysed using exploratory tests combined with simulation. Thirty points—with a priori known positions—are measured (averaging 150 repeated measurements per point) while the number of transmitters for the desired points is deliberately changed from two to eight. Coordinates position errors are determined considering the difference between the a priori known coordinates' position, and the coordinates' position of the points, calculated by triangulation. Standard deviations (σ_x, σ_y, σ_z) related to the coordinates position errors are plotted against the number of transmitters

high. To verify the potential of iGPS™, some practical experiments were carried out. In the first one, four transmitters were used to cover a relatively large working area (about 300 m^3, considering a plant layout) (Maisano et al. 2008).

The influence of the number of transmitters "seen" by a sensor on its position error was analysed, using exploratory tests combined with simulation. Thirty points—with a priori known positions—were measured (averaging 150 repeated measurements per point) while the number of transmitters for the desired points was deliberately changed from two to four transmitters. Coordinates position errors (residuals) were determined considering the difference between the a priori known position coordinates, and the position coordinates of the points, calculated by triangulation. Position errors relating to all 30 points were put together, showing a normally distributed pattern.

In the simulation experiment the effect of the number of transmitters was studied, varying the transmitters number from two to eight. The results showed a very large difference in performance between two and three transmitters. Passing from three to four transmitters, the improvement in the accuracy is still large. For five or more transmitters, improvement showed to be negligible. This behaviour is shown in Fig. 2.5; the standard deviations (σ_x, σ_y, σ_z) relating to the coordinates position errors are plotted based on the number of transmitters. We notice that the position error standard deviations with two and three transmitters are much larger than when there are four or more transmitters.

According to the results presented before, the quality of measurement is significantly influenced by the number of transmitters. For instance, during the measurement by four transmitters, if the path between a transmitter and a desired sensor is accidentally blocked and the sensor can only see three of the transmitters, the measurement quality will drop. This can happen when the line of sight between a

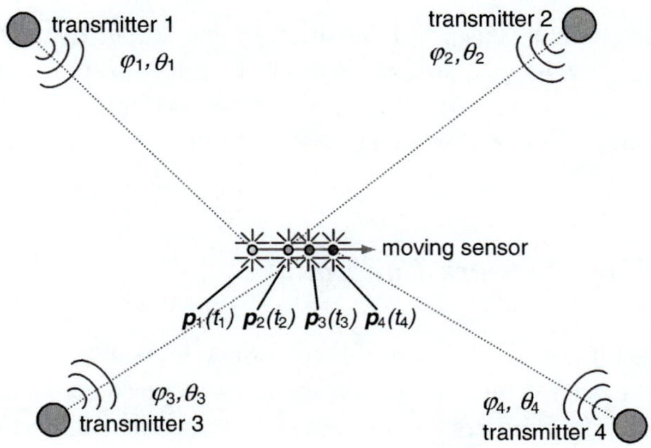

Fig. 2.6 If a sensor moves, data from transmitters is inevitably received in different instants (t_1, …, t_4). Even if the difference consists of a few tens of a second, it produces an error in the localization of the sensor. Of course, the faster the sensor moves, the larger the error becomes (Maisano et al. 2008). (with permission)

sensor and one or more transmitters is obstructed by the operator or the workpiece. Consequently, the transmitters should be arranged around the measuring area in suitable positions to gain maximum coverage (e.g., near the ceiling, to reduce the risk of obstructions). Chapter 4 reports some indications on how to arrange transmitters depending on the measured object shape and the measuring area.

2.3.2 Sensor Movement During Measurement

iGPS™ can be used to perform either static or dynamic measurements. For example, during aircraft assembly operations, it can be useful to perform measurements of moving parts to be aligned and assembled. However, the system performs best in static measurements. This is due to the localisation method used. The position of each sensor can be calculated by triangulation using the two angles (φ and θ) from each transmitter. Transmitter sampling rate depends on the angular speed of its rotating head. As explained above, spinning speed is unique for each transmitter to be differentiated. Assuming the rotation speed around 3,000 rpm, each transmitter is able to communicate with sensors about 50 times per second. Even though the differences in the transmitter sampling rate are small, it is impossible to receive concurrent data from all transmitters. The inevitable difference in data streaming is in the range of a few hundredths of a second. This effect does not create any problem for static measurements but it can affect dynamic measurements. Figure 2.6 shows such a scenario, in which sensors are moving in time (t). For any sensor, the position is calculated by triangulating data collected in very close, but different instants (t_1, …, t_4) (Moore et al. 2004).

It can be assumed, for the purpose of discussion, that data collection occurs by sensing information received firstly by transmitter 1, secondly by transmitter 2, thirdly by transmitter 3 and finally by transmitter 4. At time t_1, angles of the moving sensor with respect to transmitter 1 are read when sensor is localized in position $\boldsymbol{p}_1$, at time t_2, angles of the moving sensor with respect to transmitter 2 are read when sensor is in position p_2 and so on. Even if the difference consists of a few tenths of a second, it produces a localization error. Therefore, the faster the sensor moves, the larger the error becomes. Section 2.4.2 will give further information on dynamic measurement.

2.3.3 Network Calibration

With iGPS™ it is possible to arrange transmitters in different ways, depending on the desired measuring area and the workpiece geometry. Every time the position of the transmitters is changed, a network calibration should be performed. Obviously, this step needs to be completed before performing measurements and its accuracy has strong effects on the accuracy of the measurements results (Patwari et al. 2005). For this reason, iGPS™ software provides a semi-automated network calibration procedure that requires a few initial measurements, which can be done manually or automatically, for example by a robot. During the calibration procedure, the system scale is determined by placing two sensors at known distances within the measuring area, in at least eight different positions and orientations. To that purpose, a reference bar of a priori known length can be used.

Many reference bars with different lengths but similar uncertainties are used. Longer reference bars normally generate better results in the above-mentioned network calibration process (Zakrzewski 2003). However, the use of reference bars which are too long is not practical and may produce other errors (e.g., flexing or thermal expansion of the bar, error related to the angles uncertainty), which may inversely influence the accuracy in the localization of transmitters.

2.3.4 Environmental Factors

iGPS™, like most measuring instruments, is sensitive to several environmental factors, in particular temperature, light and vibration. It is well known that laser signals are sensitive to changes in air conditions, especially in terms of temperature, since this can undergo both temporal and spatial variations within large working volumes. Light typically has a "*go, no-go*" effect, that is to say if sensors are exposed to light, the laser beams can be "obscured" and consequently measurements cannot be performed at all. Vibrations are another source of error that can produce small movements in the measured workpiece or the measuring equipment. This effect can be large, and should be considered when analysing the results.

To filter errors from the measurement due to external factors such as light, temperature or vibrations, the iGPS™ software provides several diagnostic controls. The reliability of measurements increases significantly by using auxiliary sensors, which are placed in fixed positions at a priori known distances. With these sensors, the system can correct the initial network calibration in real-time,

Table 2.2 Results of the iGPS™ exploratory tests, performed in the specific testing conditions described in Sect. 2.4.1

Test	Repeatability			Reproducibility			Accuracy			
Mean std deviation (mm)	σ_x0.06	σ_y0.06	σ_z0.04	σ_x0.16	σ_y0.16	σ_z0.08	σ_x0.17	σ_y0.17	σ_z0.10	σ_{DIST}0.21

by compensating the changes in the environmental conditions of the measuring field, and determining whether the system is conforming to the desired tolerance (Kang and Tesar 2004).

2.4 Metrological Performance Characterization

2.4.1 Static Measurements

To evaluate the iGPS™ metrological performance in static conditions, explorative tests were performed in the Industrial Metrology Laboratory of the University of Bath. Here is a short description of the experimental setup:

- use of 4 transmitters;
- measuring area of about 60 m^2 (6 × 10 m, considering a plant layout);
- the network calibration was performed using the mobile probe as a reference bar.

Globally, iGPS™ performance was estimated through three tests:

1. *Repeatability test* In this test, a point within the working volume was measured repeatedly about 150 times to benefit from the high sampling rate of the instrument. During these measurements, the probe was left in a fixed position. The test was repeated for 30 different points in different areas of the working volume. For each point coordinate, the residuals between the individual measurements and their average value were calculated. Then, for each Cartesian coordinate (x, y, z) the residuals from all 30 points were put together. Residuals showed a normally distributed pattern. The repeatability indicator was given by the standard deviations (σ_x, σ_y, σ_z) related to each Cartesian coordinate residual (see Table 2.2).
2. *Reproducibility test* This test was similar to the previous one, the only difference being that the probe was replaced before each single point measurement. Hence, each point was approached from a different direction, using different orientations of the probe. In general, reproducibility gives a preliminary indication of the actual system's accuracy, whereas repeatability gives a preliminary indication of the potential system's accuracy, obtainable by compensating the most important causes of systematic errors. Table 2.2 shows the standard deviations related to each Cartesian coordinate. As expected, the standard deviations are higher than those of repeatability tests.

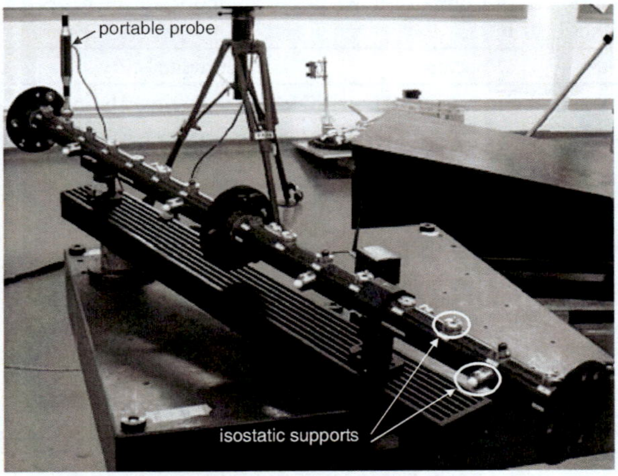

Fig. 2.7 National Physics Laboratory artefact, used for iGPS™ experiments (Cross et al. 1998; Maisano et al. 2008). (with permission)

3. *Accuracy test* This test was performed using a calibrated reference artefact with known dimensions (Cross et al. 1998). The reference artefact consisted of two bars of 1 m, assembled to create a 2-m long reference bar. The reference bar was made of composite materials with different isostatic supports on which the mobile probe can be placed during measurement (see Fig. 2.7).

The nominal dimensions of the artefact (the nominal position of reference points and the nominal distances between points) were calibrated using a laser interferometer and a CMM, which are more accurate by at least two orders of magnitude than iGPS™. These distance measurements were repeated by placing the artefact in 30 different positions and orientations within the measuring area. To reproduce a common measuring strategy, each position was calculated by averaging 150 single point measurements. The standard deviation related to the distance residuals (σ_{DIST} in Table 2.2), i.e., the differences between nominal distances and distances measured with iGPS™, was also calculated. Moreover, for each point coordinate, the residuals between the measured and the nominal position Cartesian coordinates were calculated. The standard deviations related to the coordinates (σ_x, σ_y, σ_z) are then calculated. The residuals are verified to be normally distributed. Based on these results, the iGPS™ uncertainty (referring to a $\pm\ 2\sigma$ interval) can be roughly estimated to be less than 1 mm.

The results of these exploratory tests are reported in Table 2.2. Taking into consideration different testing conditions, these results are reasonably consistent with the results of some tests carried out by iGPS™ constructors (ARC Second 2010). In general, the σ_z value is lower than σ_x and σ_y, for repeatability, reproducibility and accuracy tests. This is due to the geometric configuration of the network devices as transmitters are mounted on tripods, which are set more or less at the same height. Therefore, they can be considered to be approximately placed on a horizontal plane (*XY*) perpendicular to the vertical (*Z*) axis (Patwari et al. 2005).

2.4.2 Dynamic Measurements

Indoor GPS is a distributed metrology system that offers a relatively low cost alternative for measuring and tracking an unlimited number of moving objects in a large volume. While the static coordinate measurement performance of the iGPS™ was analysed in a fair amount of detail (Maisano et al. 2008; Muelaner et al. 2008a, b; ARC Second 2010), little existing literature gives detail on its dynamic performance (Wang et al. 2009). For this reason, it is worthwhile offering some ideas about the dynamic tracking performance of this system. The experiments described in this section attempt to analyze the dynamic repeatability of iGPS™, comparing it with the performances of a laser tracker under the same experimental conditions. Of course, consistently with what is reported in Sect. 2.3, the iGPS™ metrological performance is expected to deteriorate in dynamic conditions.

The equipment used for these exploratory experiments consisted of:

- an iGPS™ system with four transmitters and a portable probe;
- a FARO™ laser tracker with a single point angular accuracy (2σ) of $18 + 3$ μm/m, and a distance accuracy (2σ) of $20 + 1.1$ μm/m in ADM mode (Estler et al. 2002; ANSI/ASME B89.4.19-2006 2006; FARO Europe 2004);
- an industrial articulated robot—KUKA™ KR240-2—with a static point repeatability of ± 0.12 mm, and a maximum movement speed of 2 m/s. This robot was used to move the iGPS™ and the laser tracker probes controlling the trajectory and the speed of the movement.

Measurement instruments were arranged as shown in Fig. 2.8. The laser tracker is placed approximately 4 m from the robot base. The iGPS™ transmitters were arranged in a 'C' shaped network approximately 6–7 m from the robot base. The robot was programmed to run through three orthogonal linear trajectories (*X*, *Y* and *Z*). The lengths of these trajectories were approximately 1 m, 1.7 m and 0.75 m for *X*, *Y* and *Z* respectively. These lengths correspond to the full stroke values of the robot through the three axes.

The iGPS™ portable probe and the laser tracker (*cat's eye*) retroreflector were in turn attached to the robot arm, in the same position (a 38 mm steel nest). This position was defined as the Tool Centre Point (TCP) of the robot. The nominal trajectory of the TCP is known a priori and compared with the positional measurements by the iGPS™ and the laser tracker. Specifically, the robot arm was moved at six different speeds (10, 100, 300, 500, 750 and 1,000 mm/s) along the three axes (*X*, *Y* and *Z*). Data related to the *Z* trajectory were not available for 750 and 1,000 mm/s speeds, due to robot speed limitations for movements in that direction.

Position error of the robot in dynamic conditions is a one to two order of magnitude smaller than the expected error of measurement of iGPS™ and the laser tracker, thus it can be neglected (Kuka Roboter 2002).

Measurement errors can be quantified by the mean distance between the points measured by the two instruments (iGPS™ or laser tracker) and the corresponding (robot) nominal positions (see Fig. 2.9).

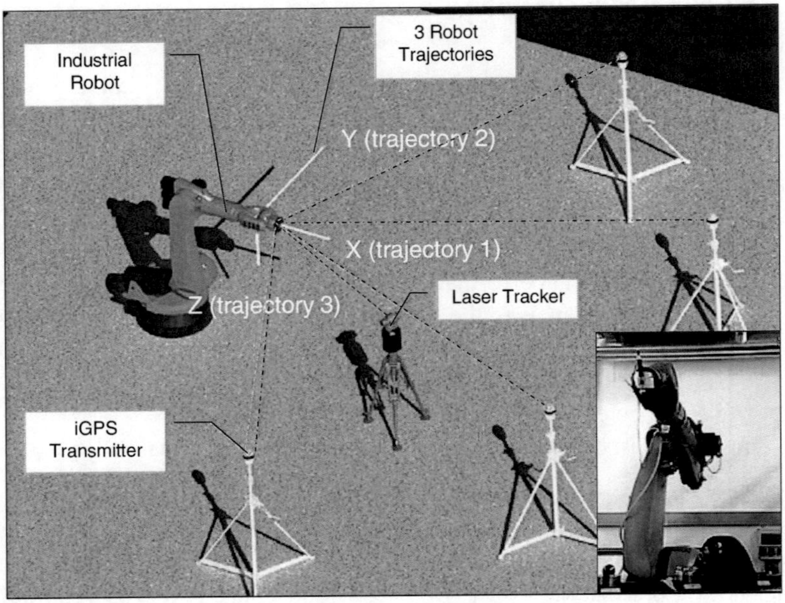

Fig. 2.8 Schematic illustration of the experiment layout and picture of the robot carrying the iGPS™ portable probe (Wang et al. 2009). (with permission)

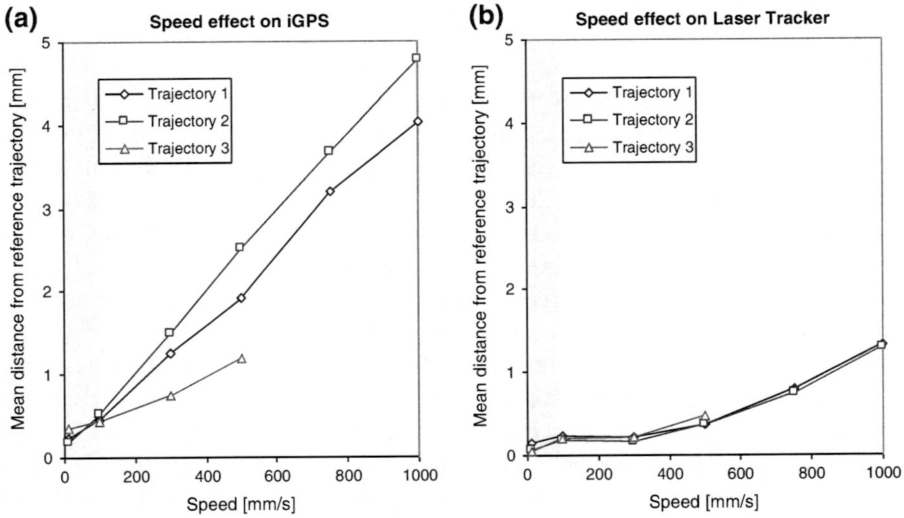

Fig. 2.9 Mean distance between the points measured by the two instruments — **a** iGPS™ and **b** laser tracker — and the corresponding (robot) nominal positions. Results obtained using six different speeds (10, 100, 300, 500, 750 and 1,000 mm/s) and three trajectories. Data related to trajectory 3 are not available for speeds of 750 and 1,000 mm/s, due to robot speed limitations for vertical movement. The *grey area* highlights the typical speed values for assembly processes (smaller than 100 mm/s)

It can be observed that, as speed increases, the repeatability of iGPS^{TM} and laser tracker tends to decrease. At low speed (say smaller than 100 mm/s) the difference between the two instruments in terms of position error is not very significant. On the other hand, when the speed of the robot arm is 1,000 mm/s, position error of iGPS^{TM} is about four times as much as that of the laser tracker.

However, this research indicates that the dynamic context in which these instruments are used is to track moving objects during assembly processes, with speeds typically lower than 100 mm/s (Rooks 2001; Alici and Shirinzadeh 2003; Eastwood et al. 2003; Webb and Eastwood 2004; Kayani and Jamshidi 2007). Therefore, in this case, both instruments' performance is similar. One advantage of iGPS^{TM} compared with laser tracker is its ability to track more than one point (usually three or more points for each of the parts to be assembled) by using just one iGPS^{TM} network.

Chapter 3
The Mobile Spatial Coordinate Measuring System

3.1 System Architecture

The Mobile Spatial coordinate Measuring System (MScMS) is a portable distributed system, designed to perform indoor dimensional measurements of large-sized objects (Franceschini et al. 2009d). The MScMS architecture consists of three basic elements (see the schematic representation in Fig. 3.1):

- a network (or "constellation") of sensing devices, distributed within the working volume;
- a portable probe to "touch" the points of interest on the surface of the measured object, so as to obtain their spatial coordinates;
- a data processing unit, running ad hoc application software, to acquire, store and elaborate measurement data.

Within the MScMS framework, two systems, based on acoustic and optical technology respectively, have been developed at the Industrial Metrology and Quality Laboratory of Politecnico di Torino—DISPEA. The first is MScMS-I, which is based on UltraSonic (US) devices evaluating relative distances by the Time-of-Flight of US signals. The second is MScMS-II, which implements a network of InfraRed (IR) cameras, tracking light emitting or light reflecting markers.

The different technology of the two systems influences (1) the procedure for calibrating the network and (2) the techniques for locating the portable probe during measurements.

A detailed description of the two systems, as to working principles, localization techniques, prototype development and metrological performance, is provided in Sects. 3.2 and 3.3.

F. Franceschini et al., *Distributed Large-Scale Dimensional Metrology*,
DOI: 10.1007/978-0-85729-543-9_3, © Springer-Verlag London Limited 2011

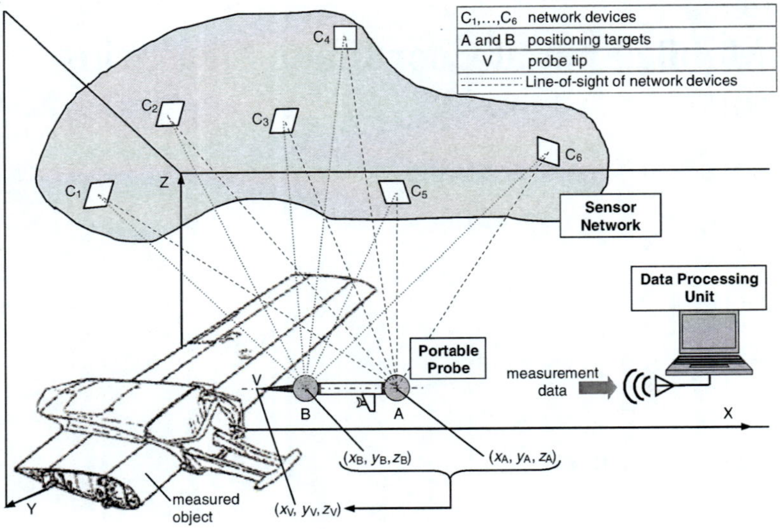

Fig. 3.1 MScMS architecture. The spatial coordinates of network devices C_i ($i = 1,\ldots,6$), positioning targets (A and B) and probe tip (V) are given in a room-aligned world coordinate reference frame with axes X, Y, and Z. *Dashed* and *dotted lines* represent line-of-sight of network devices (adapted from Franceschini et al. 2009d). (with permission)

3.1.1 Sensor Network

The network of distributed sensors is aimed at providing reference points for locating the portable probe, which is equipped with a set of positioning targets. Each sensing device, being able to establish a communication link with at least one positioning target within its "field-of-sensing", contributes to the localization of the probe. The basic operations for the network setup are:

1. configuration design: this entails network sizing, i.e., determining the number of network devices, and layout definition, i.e., rough positioning and orientation of network devices;
2. sensor deployment: this consists in placing the devices within the working volume according to the layout design;
3. network calibration: this provides actual sensor position and orientation according to self-calibration techniques (see Chap. 5 for a detailed description of calibration techniques).

It is noteworthy that the network design has to take into account technology-related issues (e.g., field-of-sensing, communication links), localization techniques, physical and environmental constraints. The aim of this phase is to determine a network layout that guarantees the coverage of a measurement region/set, i.e., able to localise each point within this region/set. A detailed analysis of sensor positioning issues is presented in Chap. 4.

According to its network architecture, the MScMS can fit working volumes of different size and shape, by varying the number of network devices. In this sense, the MScMS is scalable and modular.

3.1.2 Portable Probe

The portable probe is equipped with two positioning targets (A and B). These targets are aligned with the tip (V), which is brought into contact with the points of the measured object surface. The position of the probe's targets can be obtained using either distance or angle information from the network devices, as detailed in Sects. 3.2.2 and 3.3.3. Next, the position of V can be calculated. Since V lies on the same line of targets A and B (see Figs. 3.1, 3.2), this line can be univocally determined when the coordinates of points $A \equiv (x_A, y_A, z_A)$ and $B \equiv (x_B, y_B, z_B)$, and distances $d(A - B)$ and $d(A - V)$ are known. The parametric equation of this line is:

$$\begin{cases} x = x_A + (x_B - x_A) \cdot t \\ y = y_A + (y_B - y_A) \cdot t \\ z = z_A + (z_B - z_A) \cdot t \end{cases} \tag{3.1}$$

The distance $d(A - V)$ can be expressed as:

$$d(A - V) = \sqrt{(x_A - x_V)^2 + (y_A - y_V)^2 + (z_A - z_V)^2} \tag{3.2}$$

Coordinates of point $V \equiv (x_V, y_V, z_V)$ are univocally determined by solving a system of four equations in four unknown variables (x_V, y_V, z_V, and t_V):

$$\begin{cases} x_V = x_A + (x_B - x_A) \cdot t_V \\ y_V = y_A + (y_B - y_A) \cdot t_V \\ z_V = z_A + (z_B - z_A) \cdot t_V \\ d(A - V) = \sqrt{(x_A - x_V)^2 + (y_A - y_V)^2 + (z_A - z_V)^2} \end{cases} \tag{3.3}$$

Replacing terms x_V, y_V, z_V in the fourth equation:

$$d(A - V) = \sqrt{[(x_B - x_A) \cdot t_V]^2 + [(y_B - y_A) \cdot t_V]^2 + [(z_B - z_A) \cdot t_V]^2} \tag{3.4}$$

Then:

$$t_V = \frac{d(A - V)}{\sqrt{(x_A - x_B)^2 + (y_A - y_B)^2 + (z_A - z_B)^2}} = \frac{d(A - V)}{d(A - B)} \tag{3.5}$$

The denominator of Eq. 3.5 is the distance $d(A - B)$ between the two targets installed on the portable probe.

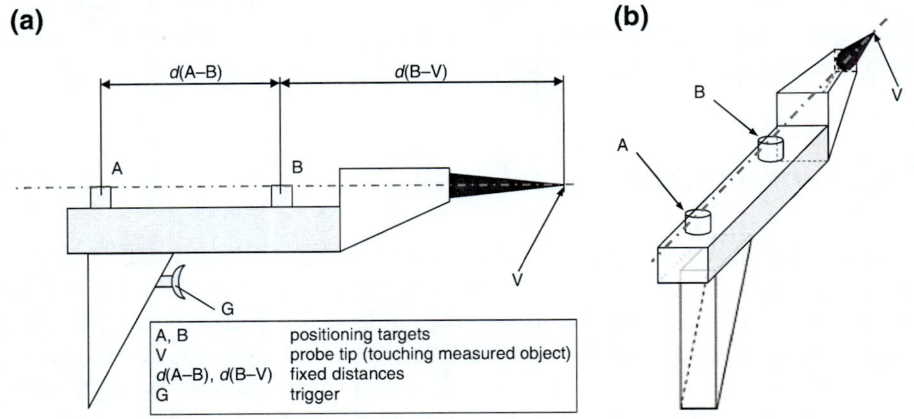

Fig. 3.2 Schematic representation of the portable probe **a** Orthogonal view **b** Perspective view (adapted from Franceschini et al. 2009d). (with permission)

In conclusion, the coordinates of the point V can be calculated as:

$$\begin{cases} x_V = x_A + (x_B - x_A) \cdot \dfrac{d(A-V)}{d(A-B)} \\ y_V = y_A + (y_B - y_A) \cdot \dfrac{d(A-V)}{d(A-B)} \\ z_V = z_A + (z_B - z_A) \cdot \dfrac{d(A-V)}{d(A-B)} \end{cases} \tag{3.6}$$

Equation 3.6 univocally locates the point V using spatial coordinates of targets A and B. Distances $d(A - B)$ and $d(A - V)$ are known a priori as they depend on the probe geometry.

The previous model is based on the assumption that A, B and V are punctiform geometric elements and the position of A and B is known. In practice, the model is inevitably approximated because elements A, B and V have non punctiform dimensions. To minimize position uncertainty for point P, the following condition should be approached: $d(B - V) \ll d(A - V)$ (Zakrzewski 2003).

3.1.3 Data Processing Unit

The data processing unit (DPU) is in charge of communicating with the sensing devices, gathering the measured data, and processing the data to provide dimensional measurement results. Depending on the network layout and the measuring hardware, different levels of interaction between sensors and processing unit could be established by using different network structures (see Fig. 3.3). Considering a flat structure (Fig. 3.3a), each sensing device (S) is linked to the DPU and transmits the measured data. On the other hand, to follow a hierarchical approach

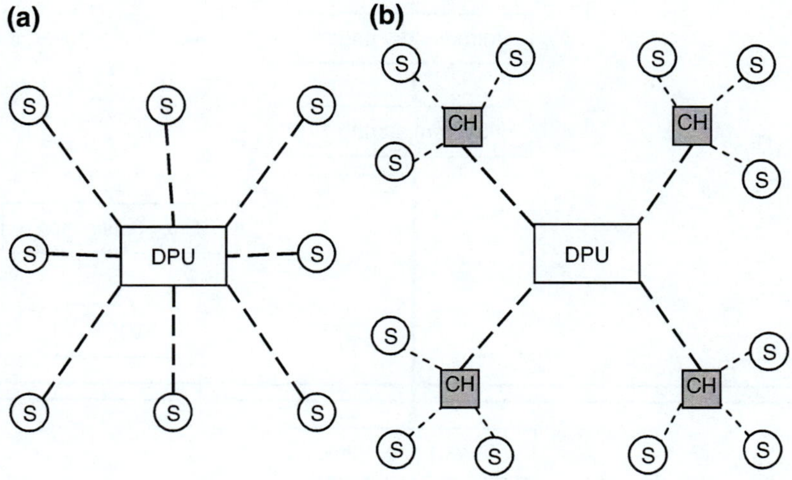

Fig. 3.3 Network layout: **a** flat structure; **b** hierarchical structure. DPU denotes a data processing unit, CH indicates a clusterhead, and S denotes a sensing device

(Fig. 3.3b), a more complex structure with clusters of nodes, in charge of a clusterhead (CH), could be designed. The clusterhead is responsible for gathering data from a subset of nodes and for routing them to a centralized unit, probably after some data processing (Cassandras and Li 2005).

The DPU, therefore, consists of one or more processing platform(s), connected to the network devices and/or to the portable probe through wireless communication links. Whenever multi-platform-based solutions need to be implemented, data exchange is managed through standard communication protocols (e.g., TCP/IP, UDP).

The software architecture is modular (see Fig. 3.4). Each module is associated with a specific activity (system startup, dimensional measurements, measurement results visualization).

Modules are linked together by different operational paths. Each path represents a sequence of screenshots. The great advantage of a modular structure is that it can be progressively extended on the basis of the system evolution.

The selection between MScMS-I and MScMS-II is made in the introductory page. The following modular structure is common to the two systems. The system startup consists in opening the communication port between the DPU and the sensor nodes in charge of transmitting measurement data. An error notification is provided in case of connection failure. The MScMS operation task management includes the following functions (see Fig. 3.4):

- *Probe qualification* this consists in identifying the portable probe according to the number of positioning targets and geometry;
- *Network calibration* this launches the calibration software application and returns the network localization data (sensor positions, orientations and other internal parameters);

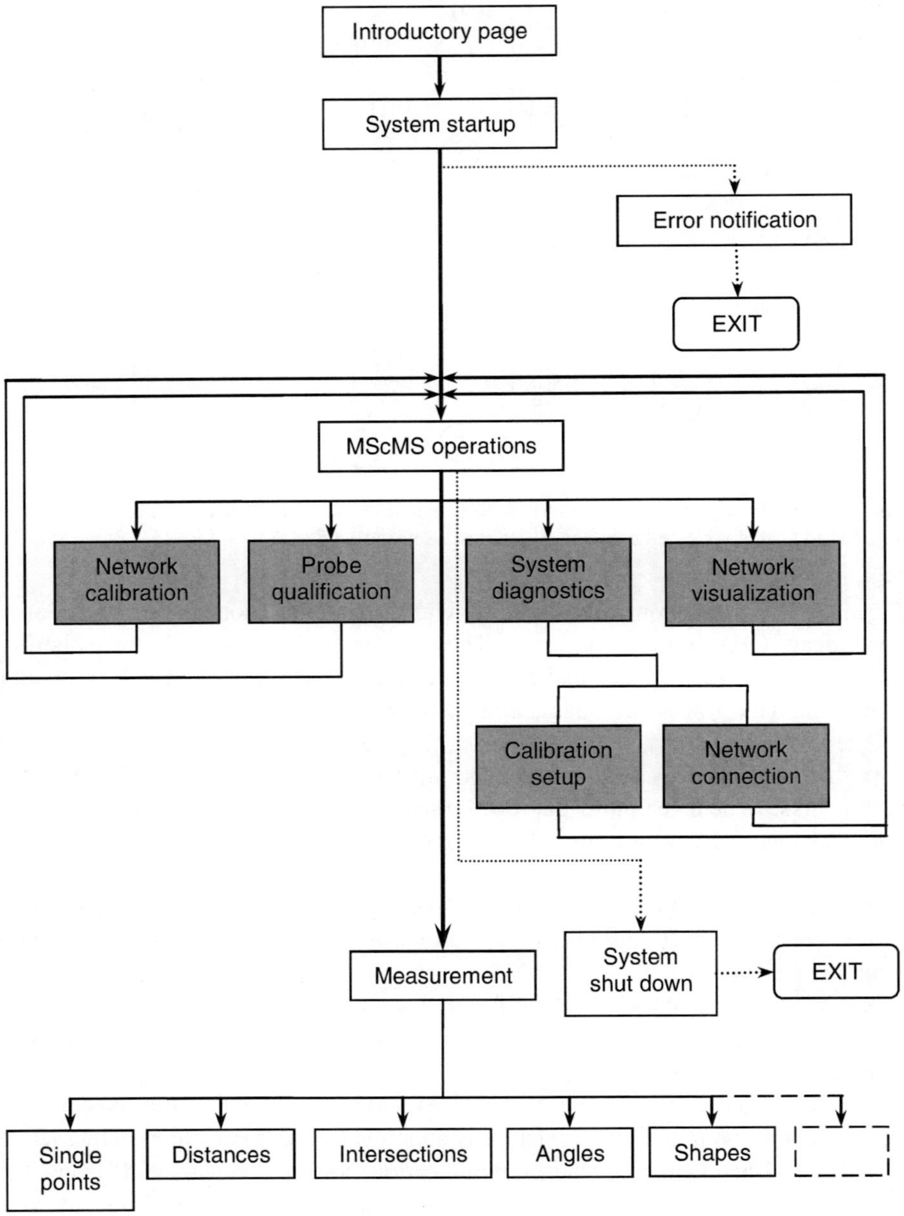

Fig. 3.4 Schematic representation of the MScMS software architecture. *Bold lines* indicate the basic path for running a measurement session with a previously calibrated network and a qualified probe. *Dotted lines* identify paths entailing the graphical user interface exiting. *Grey boxes* represent system presetting operations, generally carried out before the measurement session. *Dashed lines* indicate possible additional paths related to further system capabilities

- *System diagnostics* this consists in checking the validity of available network localization data by measuring a reference calibrated artefact (calibration setup) and/or in checking the network communications links (network connection);
- *Network visualization* this launches an application for visualizing the network layout and the actual working volume, according to sensor positioning and field-of-sensing;
- *Measurement* this represents the main interface, used to perform dimensional measurements. It links to functions similar to those typically implemented by CMM software packages. MScMS, likewise CMMs, makes it possible to determine the shape/geometry of objects (circumferences, cylinders, plans, cones, spheres etc.), on the basis of a set of measured surface points gathered from the portable-probe, using classical optimization algorithms (Overmars 1997).

3.2 The MScMS-I: A System Based on Ultrasound Technology

The MScMS-I portable probe is equipped with two wireless devices, identical to those making up the network. These devices, known as Crickets , were developed by Massachusetts Institute of Technology and manufactured by Crossbow Technology (2010). Cricket devices are equipped with RadioFrequency (RF) and UltraSonic transceivers. Working frequencies are 433 MHz (on RF) and 40 kHz (on US). Each device uses an Atmega 128L microcontroller operating at 7.4 MHz, with 8 kB of RAM, 128 kB of FLASH ROM (program memory), and 4 kB of EEPROM (as mostly read-only memory). Power is provided by two "AA" batteries of 1.5 V (Balakrishnan et al. 2003).

Cricket devices are characterized by small dimensions (see Fig. 3.5), easy to move, and cheap (a mass-produced unit cost of €10–€20). These characteristics make them optimal for ad hoc wireless sensor network applications (Priyantha et al. 2000).

While the communication of RF sensors is almost omnidirectional and may be up to 25 m, the communication range of US sensors is limited by a "field-of-sensing" schematically represented by a cone with opening angle of about 80° and range of no more than 6 m (see Fig. 3.6).

Signal strength outside the cone drops to 1% of the maximum value (Priyantha et al. 2000). Therefore, it is important to provide full coverage to the area served by network devices by positioning the US transmitters within the measurement volume. In general, the most practical solution is to mount them on the ceiling or at the top of the measuring volume, as shown in Fig. 3.6. To increase the working volume coverage it is necessary to increase the number of network devices. This problem is discussed in depth in Chap. 4. An exhaustive description of the Cricket US transceivers is presented in the first part of Chap. 7.

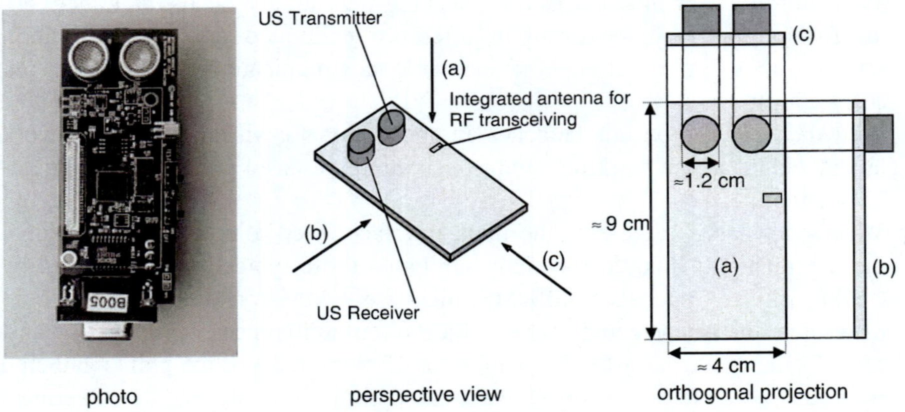

Fig. 3.5 A Cricket device manufactured by Crossbow Technology (Crossbow Technology 2010). (with permission)

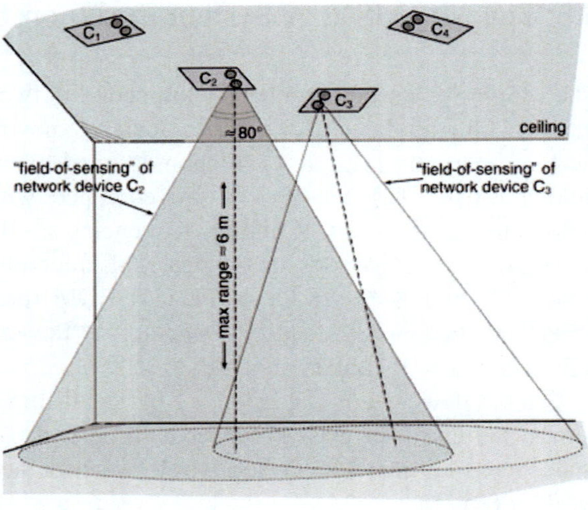

Fig. 3.6 Representation of the "field-of-sensing" associated with US network devices (adapted from Franceschini et al. 2009d). (with permission)

The system makes it possible to calculate the position—in terms of spatial coordinates—of the object points that are "touched" by the probe. US transceivers are used to evaluate mutual distances among different Cricket devices, while RF transceivers are used to send and receive this distance information to and from neighbouring devices. More precisely, when a trigger mounted on the portable probe is pulled, data is sent to the DPU via Bluetooth. Then it is used to calculate the current coordinates of the probe tip and, in turn, for different types of processing (reconstruction of curves, or surfaces of the measured objects).

Network devices (Crickets) operate as reference points, or beacons, for the portable probe. The spatial localisation of the network devices follows a semi-automatic procedure, which will be the object of Chap. 5.

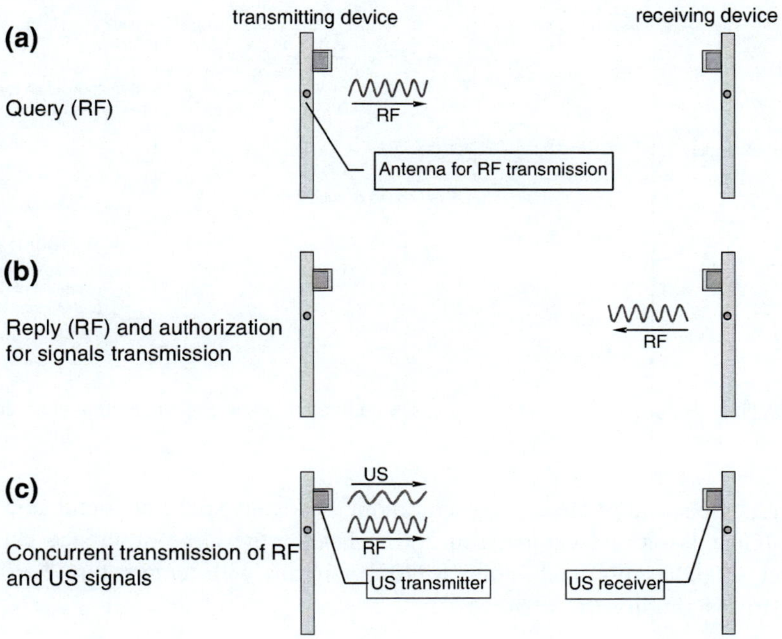

Fig. 3.7 Communication scheme implemented by Cricket devices (Priyantha et al. 2000). (with permission)

3.2.1 Working Principles

The technique, implemented by each pair of Crickets to estimate mutual distance, is known as Time Difference of Arrival (TDoA). It is based on the comparison between the propagation time of two signals with different speeds (RF and US in this case) (Savvides et al. 2001). TDoA technique is described as follows:

- at random time intervals, included between 150 and 350 ms, each device transmits an RF query-packet to other devices within its communication range, checking if neighbouring Crickets are ready to receive an US signal (Fig. 3.7a) (Priyantha et al. 2000);
- ready devices reply sending an RF acknowledgement authorizing the next signal transmissions (Fig. 3.7b);
- the querying Cricket is now authorized to concurrently send an RF and a US signal (Fig. 3.7c);
- the receiving devices measure the time lapse between the reception of the RF and the US signals (see Fig. 3.8).

The distance between the two devices is calculated by the following formula:

$$d = \frac{\Delta t}{\frac{1}{s} - \frac{1}{c}} \tag{3.7}$$

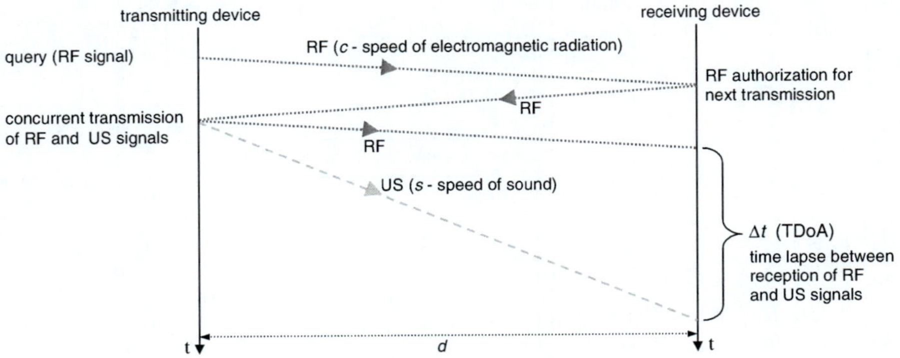

Fig. 3.8 Time evolution of RF and US signals: a qualitative scheme (Franceschini et al. 2009d). (with permission)

where c is the speed of electromagnetic radiations, s the speed of sound, and Δt is TDoA (Gustafsson and Gunnarsson 2003). Due to the large difference between c (about 300,000 km/s) and s (about 340 m/s in air, with temperature $T = 20°\text{C}$ and relative humidity $RH = 50\%$):

$$d \approx s \cdot \Delta t \tag{3.8}$$

Firmware is essential to organize RF and US communication among Cricket devices. It is written in NesC language, and works under the operating system TinyOS (2010). NesC is derived from C and is currently utilized to program MICA Mote devices (produced by Crossbow Technology), from which Crickets are derived. NesC is an *object-oriented* and *event-based* programming language (Gay et al. 2003). Programs are organized in independent modules, interrelated by means of reciprocal queries and replies (MIT C.S.A.I.L. 2004; Moore et al. 2004).

Figure 3.9 shows a schematic flow-chart of the Cricket firmware.

Each Cricket device performs two types of operations:

- Time-of-Flight measurement of the US signals transmitted or received from other devices. At random time intervals, included between 150 and 350 ms, each device tries to synchronize itself with neighbours, in order to exchange US signals. Synchronization information is transmitted through RF packets;
- when a Cricket receives a new distance, from a neighbour or directly measured, it stores and sends it to its neighbours by an RF packet containing a new list of inter-node distances.

Firmware coordinates communications among Cricket devices, making them able to cooperate and share information about inter-node distances. When the user pulls the portable probe trigger, all information is sent (via Bluetooth) to the DPU for processing.

Cricket devices build a wireless network of cooperating sensor nodes. To preserve network scalability, that is to make sure that the amount of information stored by each node is independent of the network size (in terms of number

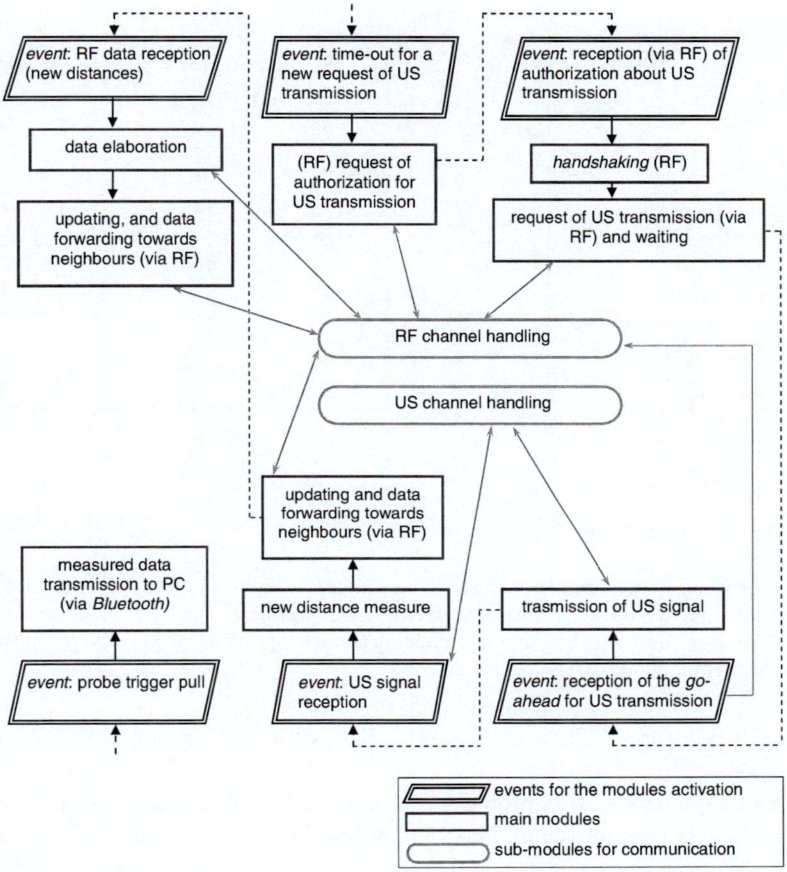

Fig. 3.9 A schematic flow-chart of the Cricket real-time firmware (adapted from Franceschini et al. 2009d). (with permission)

of nodes), each node stores the distances from the direct neighbours, i.e., those contained in its communication range (see Fig. 3.10).

3.2.2 Localisation Algorithms

The two-step localisation procedure is implemented to localise Crickets mounted on the portable probe and, subsequently, to locate the probe tip

Step 1: Localisation of Crickets Mounted on the Portable Probe
Spatial localisation of the Crickets mounted on the probe is performed by trilateration. Trilateration uses the known locations of reference devices. At least three reference points are generally needed to determine the relative location of a

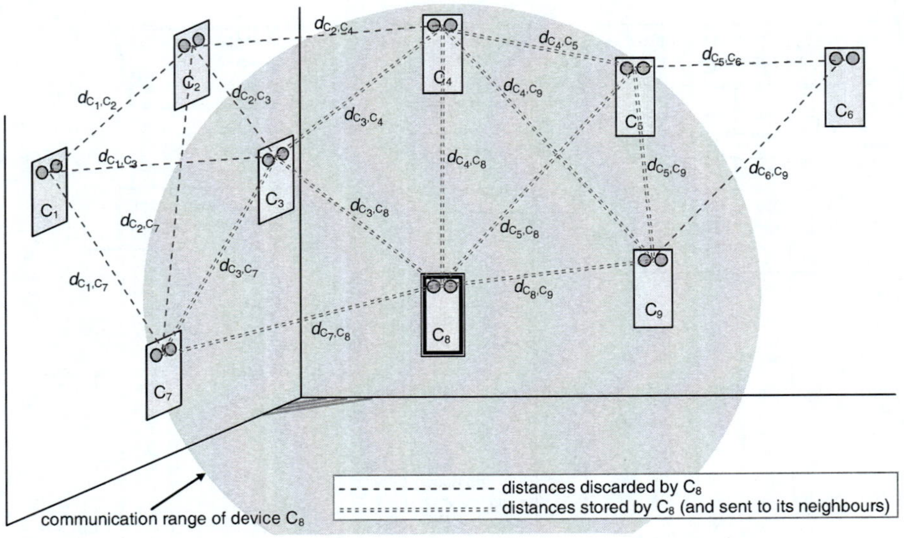

Fig. 3.10 Distance information handled by a single device (C_8). The generic term $d_{Ci,Cj}$ represents the measured distance between device C_i and device C_j (with $i = 1,\ldots,8$; $j = 1,\ldots,8$; $i \neq j$). The *grey region* highlights the communication range of the device C_8 (adapted from Franceschini et al. 2009d). (with permission)

point in a 3D space (Chen et al. 2003; Sandwith and Predmore 2001; Akcan et al. 2006).

In general, a trilateration problem can be formulated as follows. Given a set of n nodes (network devices) with known coordinates (x_i, y_i, z_i, being $i = 1,\ldots,n$) and a set of measured distances d_i, a system of equations can be solved to calculate the unknown position of a generic point $P \equiv (x_P, y_P, z_P)$ (see Fig. 3.11).

$$\begin{bmatrix} (x_1 - x_P)^2 + (y_1 - y_P)^2 + (z_1 - z_P)^2 \\ (x_2 - x_P)^2 + (y_2 - y_P)^2 + (z_2 - z_P)^2 \\ \vdots \\ (x_n - x_P)^2 + (y_n - y_P)^2 + (z_n - z_P)^2 \end{bmatrix} = \begin{bmatrix} d_1^2 \\ d_2^2 \\ \vdots \\ d_n^2 \end{bmatrix} \tag{3.9}$$

If the trilateration problem is over defined (i.e., at least 4 reference points are within the communication range of device P), it can be solved using a least-squares method (Savvides et al. 2001; Martin et al. 2002). Each unknown node (generically denoted with P) estimates its position by performing the iterative minimization of an Error Function (EF), defined as follows:

$$EF = \frac{\sum_{i=1}^{n} [d_i - G_i]^2}{n} \tag{3.10}$$

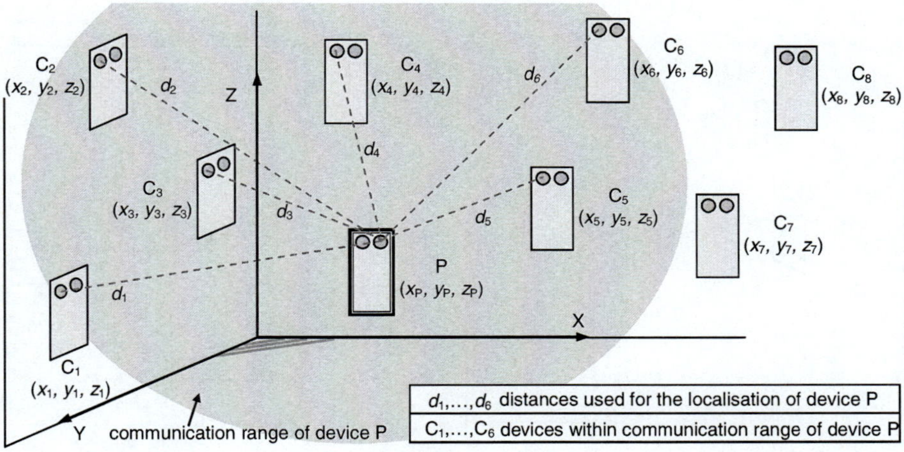

Fig. 3.11 Localisation of a generic Cricket device P ≡ (x_P, y_P, z_P) by a network of devices C_i ≡ (x_i, y_i, z_i) (i = 1,...,6). *Dashed lines* indicate the line-of-sight between device P and each network device within its communication range (*grey region*). The measured distance between the device P to locate and the generic network device C_i is indicated by d_j (j = 1,...,6). The spatial point coordinates are given in a world coordinate reference frame with axes X, Y and Z (adapted from Franceschini et al. 2009d). (with permission)

where d_i is the measured distance between the ith node and the unknown device P, G_i is the calculated Euclidean distance between the estimated position of P ≡ (x_P, y_P, z_P) and the known position of the ith device C_i ≡ (x_i, y_i, z_i), and n is the number of network devices C_i (i = 1,...,n) within the communication range of the node P.

Step 2: Localisation of Points Touched by the Probe Tip
As anticipated in Sect. 3.1, the position of the point (V) touched by the probe tip is calculated by means of Eq. 3.6, using the coordinates of Crickets A and B and the a priori known distances d(A − B) and d(A − V) (see Fig. 3.12).

It should be noted that, according to network architectures shown in Fig. 3.3, the distributed network devices are sensing nodes whereas the probe device A acts as a clusterhead. As a matter of fact, it is in charge of gathering the distance information from the network and routing them to the DPU.

3.2.3 Data Processing and Elaboration

MScMS-I data processing includes several activities:

- communication and data sharing among Cricket devices;
- data processing to determine the location of the Crickets mounted on the portable probe;
- data processing to determine the location of points touched by the probe tip;

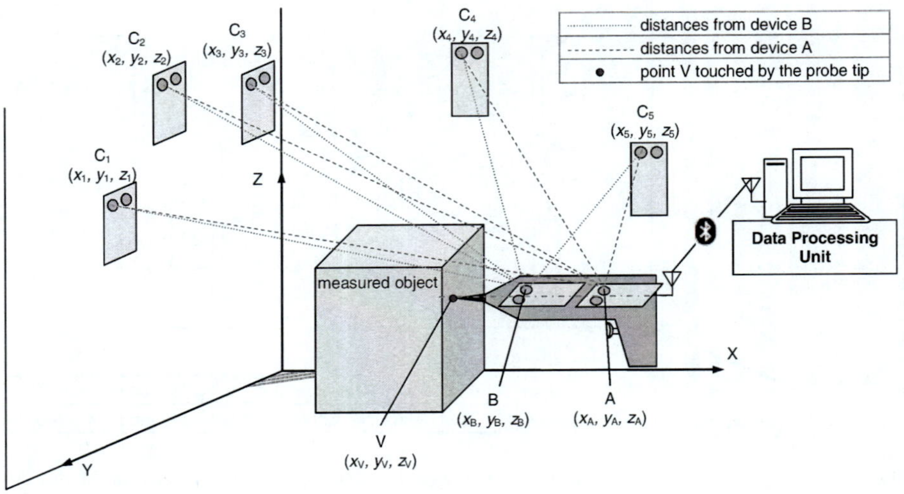

Fig. 3.12 Diagram of the procedure for localising the point touched by the probe tip. The *dashed* and *dotted lines* represent communication links between the US sensing devices of the network ($C_1,\ldots,C_5$) and the reference devices A and B, respectively. The solid line represents the Bluetooth connection that is established between the reference device A equipping the probe and the data processing unit. Through this wireless link the device A sends to the processing unit the measured distances of reference devices with respect to the network devices. Firstly the Crickets A and B, equipping the portable probe, are localised with respect to a world coordinate reference system with axes X, Y and Z (Step 1). The spatial coordinates of the probe tip V are then calculated (Step 2) (adapted from Franceschini et al. 2009d). (with permission)

- data processing to determine the shape/geometry of measured objects, on the basis of a set of points gathered from the portable probe;
- data processing related to the semi-automatic location of network devices (see Chap. 5).

The first operation is *distributed*, that is to say performed by Cricket devices. The four remaining operations are implemented by the *centralised* ad hoc software running on a PC, on the basis of the information received by the network's and probe's Crickets.

3.2.4 MScMS-I Prototype Description

Hereafter follows a brief description of the first prototype of MScMS-I.

3.2.4.1 Cricket Network

Some Cricket devices have been freely distributed around a relative large measuring area (volume of several tens of cubic meters). To make their positioning

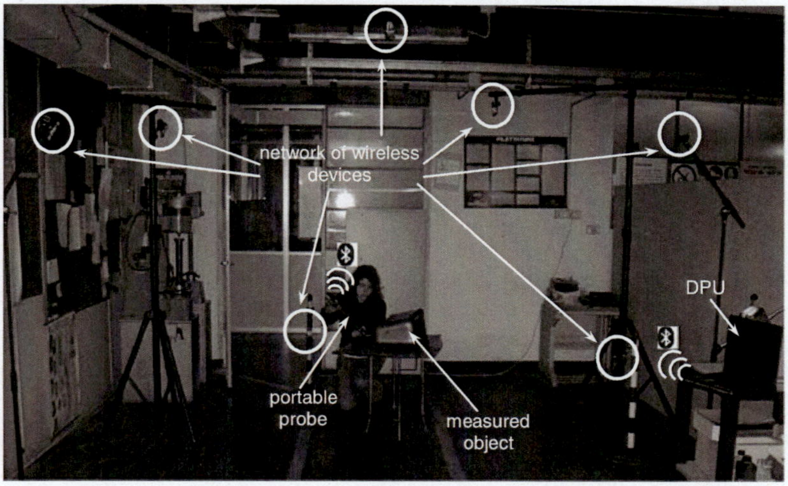

Fig. 3.13 Practical application of MScMS-I (Franceschini et al. 2009d). (with permission)

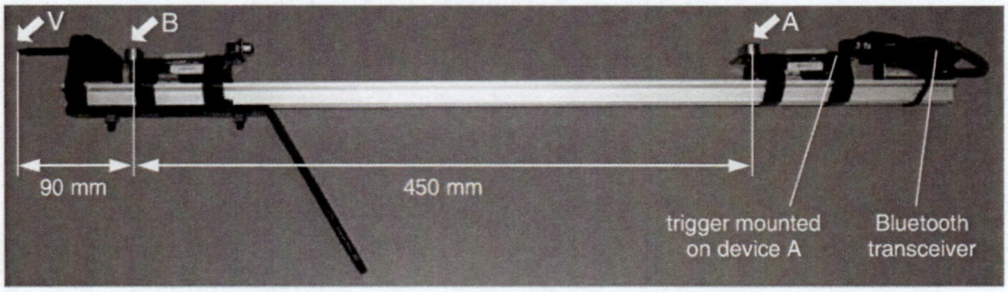

Fig. 3.14 Portable probe prototype of the MScMS-I (Franceschini et al. 2009d). (with permission)

easy, different supports are used, such as booms, articulated arms and tripods (see Fig. 3.13).

3.2.4.2 Portable Probe

This is a rigid structure which contains the following elements:

- two Cricket devices (A and B);
- a tip to "touch" the points of measured objects. Tip (V) and Cricket devices are aligned and spaced as indicated: $d(A - B) = 450$ mm and $d(A - V) = 540$ mm (see Fig. 3.14);
- a Bluetooth transceiver connected with one of the two Cricket devices, by a RS232 serial port;
- A trigger, mounted on device A, to send measurement information to a PC for centralised data processing.

3.2.4.3 Application Software

An ad hoc application software runs on a 2.5 GHz computer platform. To receive data sent by the probe, the PC is equipped with a Bluetooth transceiver. The purpose of this software is to drive the user through measurements and to make results display efficient. In more detail, the software is organized into three application modules to assist the user in the following operations:

- *Initialization* This is a guided procedure to switch on wireless devices (Crickets and Bluetooth adapter) and open the PC connection for data reception from the portable probe.
- *Semi-automatic localization of the network* This procedure will be described in Chap. 5.
- *Measurements* These include different kinds of measurement: single point measurements, distance measurements, curves and surfaces evaluation.

When the probe trigger is pulled, the application software calculates the Cartesian coordinates of the point touched by the probe tip. If measurements are taken correctly—that is to say if some of the real-time diagnostic checks described in Chap. 6 are passed—an acoustic signal is emitted. Measurement results are displayed using both numeric and graphical representations.

3.2.5 Metrological Performance Characterization

The preliminary prototype of MScMS-I was setup and tested, with the purpose of verifying system feasibility and evaluating its performance. The prototype actual performance was estimated carrying out two practical tests:

Repeatability test According to the definition of repeatability given by the International Vocabulary of Metrology (JCGM 200:2008 2008) and reported in Chap. 1, a single point within the working volume was measured and the measurement was repeated about 50 times, leaving the portable probe in a fixed position (see Fig. 3.15a). The test was repeated measuring at least 20 different points in different areas of the working volume. For each point, the standard deviations (σ_x, σ_y, σ_z) related to the corresponding Cartesian coordinates (x, y, z) were calculated.

Reproducibility test This test was carried out according to the definition of reproducibility given by the International Vocabulary of Metrology (JCGM 200:2008 2008) and reported in Chap. 1. The test procedure was similar to the previous one; the only difference is that the orientation of the portable probe was changed before each measurement, with the aim of approaching each (single) point from a different direction (see Fig. 3.15b). Precisely 20 points in different areas of the working volume were measured, changing the orientation about 50 times per point.

The reproducibility range is an index of instrument actual accuracy, whereas the repeatability variation range is an index of target instrument accuracy and aims to compensate for the most important causes of systematic errors.

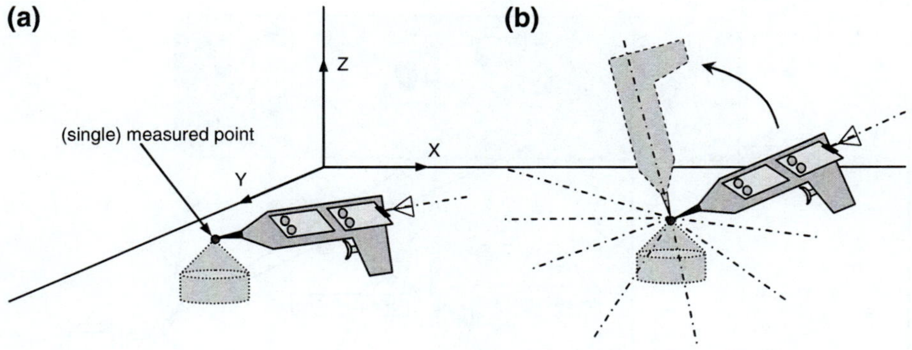

Fig. 3.15 Diagram of the practical tests carried out to evaluate MScMS-I performance. **a** Repeatability test: the position of the portable probe is fixed during measurements. **b** Reproducibility test: the direction of the portable probe is changed before every measurement (Franceschini et al. 2009d). (with permission)

Table 3.1 Results of the exploratory tests carried out on a prototype of the MScMS-I

Test	Repeatability			Reproducibility		
Mean standard deviation	σ_x	σ_y	σ_z	σ_x	σ_y	σ_z
(mm)	4.8	5.1	3.5	7.3	7.8	4.1

The statistical results of these exploratory tests are reported in Table 3.1. It is of interest that the σ_z value was basically lower than σ_x and σ_y, both in repeatability and reproducibility tests. This behaviour is due to the geometric configuration of the network of Crickets in the prototype implementation. Since network devices were mounted on the ceiling or at the top of the measuring area, they could be considered as approximately placed on a plane (XY) perpendicular to the vertical (Z) axis (see Figs. 3.6, 3.13). It was verified experimentally that the distribution of the point coordinates is approximately normal, both for repeatability and reproducibility data.

Within the context of a performance comparison with other, industrially available, metrological solutions, these results should be evaluated taking into account that they refer to a prototype implementation of the system. It is worthy to observed that system performance are, in fact, strongly affected by the sensing hardware, the layout geometry and the network sizing. The most critical aspects of the whole measuring system are a result of US sensors. In particular:

- dimensions of US transceivers;
- different types of noise affecting US signals;
- speed of sound dependence on environmental conditions;
- working volume discontinuities;
- use of amplitude threshold detection at receivers.

These aspects will be discussed individually in Chap. 7.

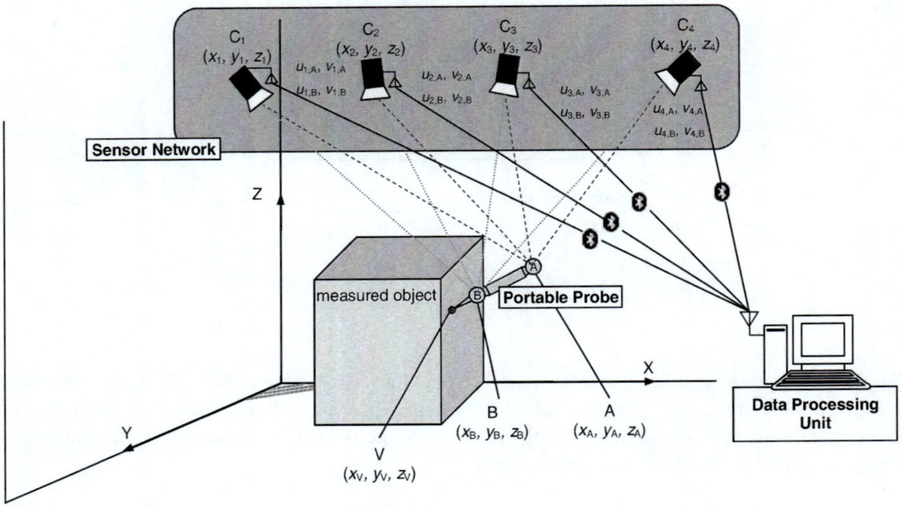

Fig. 3.16 MScMS-II architecture. The *dashed and dotted lines* represent visual links between the optical sensing devices (C_1,...,C_4) and the reference targets (indicated as A and B) equipping the portable probe. The *solid lines* represent the Bluetooth connection that is established between each node and the data processing unit. Through this wireless link each camera sends the 2D coordinates of visible markers ($u_{i,j}$, $v_{i,j}$) ($i = 1,\ldots,4$ and $j =$ A, B) to the processing unit. Markers A and B, equipping the portable probe, are localised with respect to a world coordinate reference system with axes X, Y and Z. The spatial coordinates of the probe tip V are then calculated (adapted from Galetto et al. 2011). (with permission)

3.3 The MScMS-II: An IR-based Distributed Metrological System

The MScMS-II is an indoor coordinate measuring system which exploits distributed IR optical technology. As MScMS-I the system consists of three basic units (Fig. 3.16): a network of wireless IR cameras (suitably distributed within the measurement volume to estimate 3D coordinates of reference positioning targets), a portable wireless and armless probe (equipped with two targets, to "touch" the measurement points), a data processing system (using Bluetooth connection to acquire and elaborate data sent by each network node).

3.3.1 Working Principles

The distributed network consists of wireless optical sensor devices, each one able to establish a visual link with the optical markers that are visible in its "field-of-sensing". Several types of optical sensors are available for performing both active and passive tracking. This camera-based system, working in the IR spectrum, can

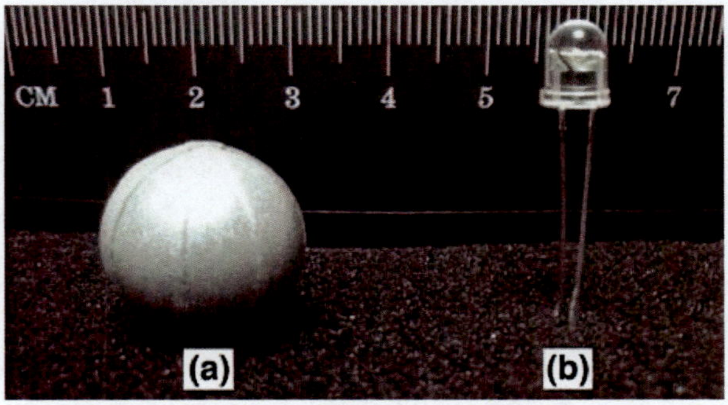

Fig. 3.17 Basic units of active and passive markers equipping the portable probe in the camera-based system: **a** a retro-reflective passive sphere; **b** an IR LED

be used to track light emitting sources (active tracking) as well as retro-reflective markers (passive tracking) (Fig. 3.17).

The active tracking is based on a mobile light source, consisting of a set of IR Light Emitting Diodes (LEDs), which have their own energy supply system and are encapsulated within a sphere made of light scattering material. On the contrary, passive tracking entails a remote fixed light source (e.g., a LED array), coupled with the camera device (see Fig. 3.25), to properly floodlight the working volume and hence the mobile spherical reflective markers.

A generic camera device is characterized by its focal length l_f and the angular Field Of View (FOV). The angular FOV is generally denoted as $(2\alpha_H \times 2\alpha_V)$, where α_H and α_V represent the half-angles of view in the horizontal and vertical plane respectively (see Fig. 3.18). On the other hand, the linear FOV is defined as that part of the scene that is visible through the camera at fixed position and orientation in the 3D space. Whereas the linear FOV, which is measured in linear dimensions (e.g., pixel units, millimetres), increases with distance, the angular FOV is constant and it is usually used as a specification parameter.

The camera position is identified by the spatial coordinates of the camera projection centre $C \equiv (x_C, y_C, z_C)$ in a world coordinate reference frame $F_{WCS}(X_{WCS}Y_{WCS}Z_{WCS})$. Each spherical light source/marker M, whose centre is located at coordinates (x_M, y_M, z_M) in F_{WCS}, appears as a bright spot in the image captured at a generic time instant by a given camera device. This corresponds to a 2D point $P \equiv (u_P, v_P)$ in the camera projection plane π, located at distance l_f from the camera projection centre along the optical axis. The 2D pixel coordinates are given in an image coordinate reference frame $F_{ICS}(UV)$, having origin in the principal point PP and axis lines U and V.

According to the available technology, a network of sensing devices provided with onboard computational capabilities has been implemented. Hence, each device can analyse the image and calculate the position coordinates of the bright spot(s) in its projection plane (NaturalPoint 2010; Vicon 2010). This information is then sent to the DPU, which uses it to reconstruct the three-dimensional position of any light source/marker by applying triangulation algorithms. Depending on the probe geometry and the marker disposition, known a priori, the 3D coordinates of the point(s) touched by the probe tip can be calculated (see Sect. 3.3.3).

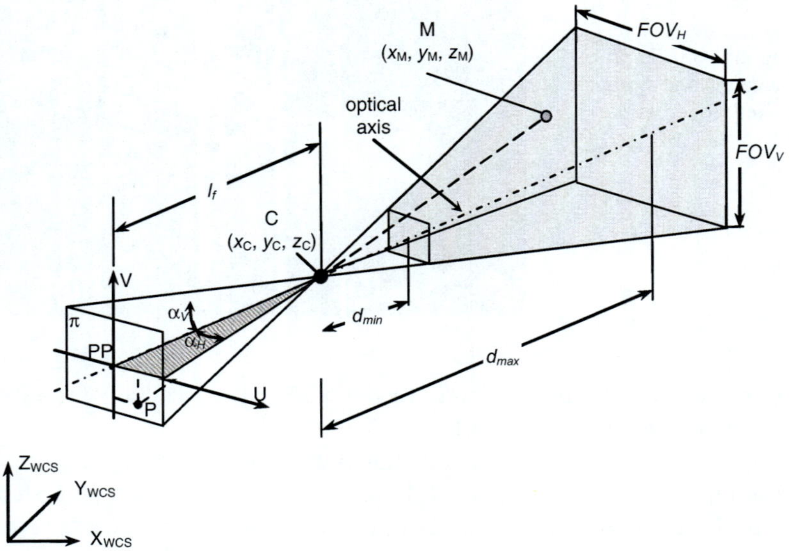

Fig. 3.18 Geometry and working principles of a camera device. The half-angles of view α_H and α_V identify the angular FOV in the *horizontal* and *vertical planes* respectively. The linear FOV in the *horizontal* and *vertical direction* is indicated with FOV_H and FOV_V, respectively. The *light grey region* represents the camera viewing volume or "field-of-sensing", i.e., the physical volume within which a retro-reflective marker is visible and traceable by the camera device. This volume, whose dimensions depend on the sensing hardware and the size of markers, is defined according to a minimum distance $d_{\min}$, related to tracking capabilities, and a maximum distance $d_{\max}$, related to visibility capabilities. It should be noted that the 3D marker position M and the camera projection centre C are defined according to a room-aligned world coordinate reference system $F_{WCS}(X_{WCS}Y_{WCS}Z_{WCS})$, whereas the 2D marker projection P refers to a local image coordinate system F_{ICS}, which originates in PP and has axis lines U and V. The origin PP is the principal point, i.e., the intersection between the optical axis and the projection plane π

3.3.2 Data Processing and Elaboration

Data processing hardware is done by a 2.5 GHz computer platform, connected to a set of IR cameras via a radio link. Providing that each camera determines the 2D coordinates of the IR spot(s) in its view plane, the embedded real-time tracking capabilities save computational effort for performing the image analysis and spot coordinates identification by the computer platform. Depending on the hardware-software configuration, a maximum number of IR sensors can be managed by a single PC unit. A modular approach, based on multiple processing units sharing the information of different camera sets, was implemented to enlarge the working volumes.

As mentioned in the discussion on software architecture in Sect. 3.1, the processing unit manages layout evaluation, system calibration, 3D point localization and data processing procedures. According to a given network configuration, the calibration block implements a camera self-calibration algorithm

(Svoboda et al. 2005). Taking as input from the camera tracking engine the 2D position estimates of a single reflective marker (calibration marker), randomly moved in k unknown positions within the working volume, it provides camera positions and orientations as well as camera internal parameters (such as focal length, coordinates of image centre, and parameters for the lens distortion model). Locating a calibrated artefact (calibration square) positioned at unknown coordinates as long as they are within the working volume, the calibration tool performs camera alignment and scaling to a user-defined coordinate reference system (see Fig. 3.19). This information is then used by the localization algorithm to perform 3D reconstruction of measurement points, according to digital photogrammetry principles. The data processing software tool, which has been designed as to accomplish system flexibility, provides capabilities to perform single point coordinate measurements, distance measurements as well as geometry reconstruction.

3.3.3 Localisation Algorithms

The 3D coordinates of the generic marker M_j can be related, through the collinearity equation, to the 2D coordinates of the corresponding image point $P_{i,j}$ onto the ith camera projection plane as functions of camera technical parameters, position and orientation. The collinearity equations represent the geometric condition under which the 3D point M_j, its projection $P_{i,j}$ onto the camera image plane, and the camera projection centre C_i lie on the same straight line (see Fig. 3.20). These equations can be written in the following form (Mikhail et al. 2001; Luhmann et al. 2006):

$$u_{i,j} = u_{0_i} + l_f c_{u_i} \left(\frac{r_{1,1,i}(x_{M_j} - x_{C_i}) + r_{1,2,i}(y_{M_j} - y_{C_i}) + r_{1,3,i}(z_{M_j} - z_{C_i})}{r_{3,1,i}(x_{M_j} - x_{C_i}) + r_{3,2,i}(y_{M_j} - y_{C_i}) + r_{3,3,i}(z_{M_j} - z_{C_i})} + \delta u_{i,j} \right)$$

$$v_{i,j} = v_{0_i} + l_f c_{v_i} \left(\frac{r_{2,1,i}(x_{M_j} - x_{C_i}) + r_{2,2,i}(y_{M_j} - y_{C_i}) + r_{2,3,i}(z_{M_j} - z_{C_i})}{r_{3,1,i}(x_{M_j} - x_{C_i}) + r_{3,2,i}(y_{M_j} - y_{C_i}) + r_{3,3,i}(z_{M_j} - z_{C_i})} + \delta v_{i,j} \right) \tag{3.11}$$

where (see Fig. 3.20):

- (x_{Mj}, y_{Mj}, z_{Mj}) are the 3D coordinates of the marker M_j in the world coordinate reference frame F_{WCS} with axes X_{WCS}, Y_{WCS}, and Z_{WCS};
- (x_{Ci}, y_{Ci}, z_{Ci}) are the 3D coordinates of the projection centre of camera C_i in the world coordinate reference frame F_{WCS};
- (u_{0i}, v_{0i}) are the coordinates of the image centre, i.e., the projection of point C_i on the camera image plane, in the image coordinate reference frame F_{ICS} with axes U and V;
- c_{ui} and c_{vi} are camera-related coefficients, needed to convert the focal length l_{fi}, from metric units to pixels, in the horizontal and vertical direction, respectively.

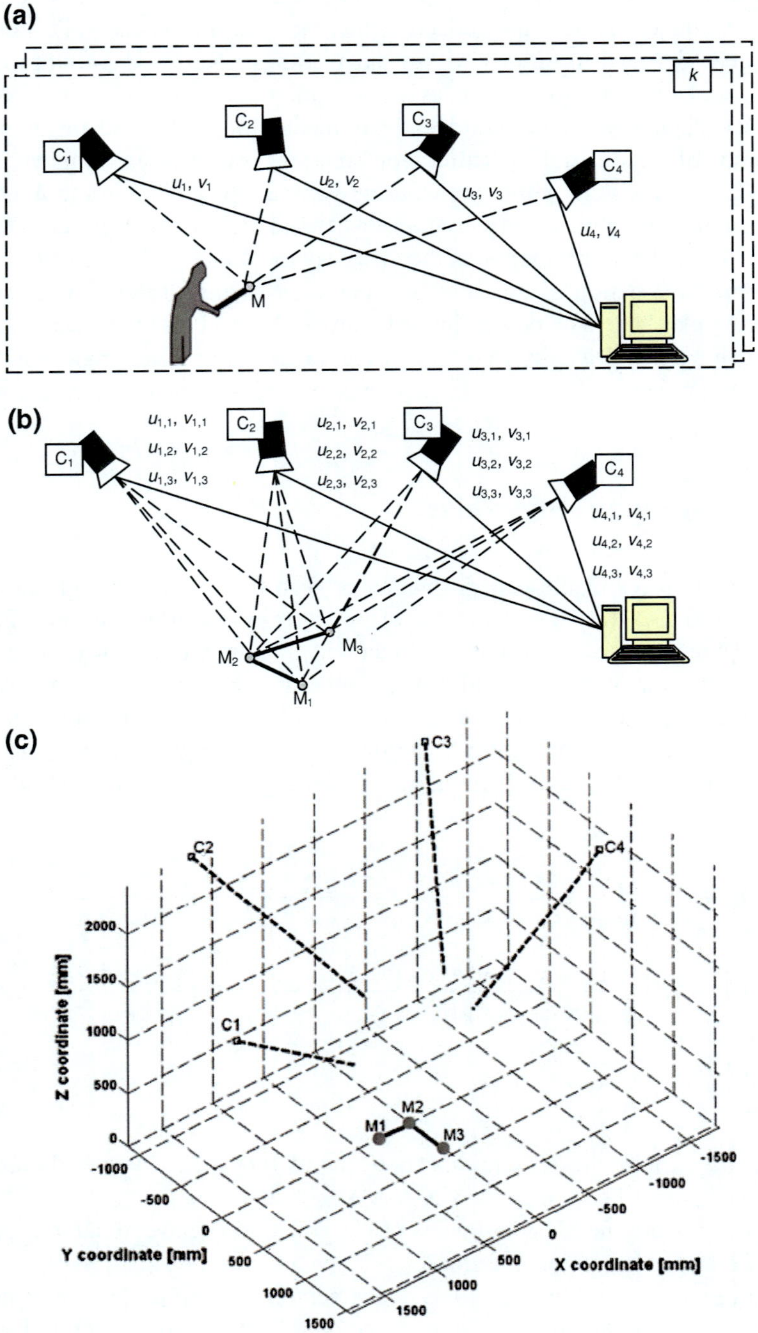

Fig. 3.19 MScMS-II network calibration procedure. **a** Calibration data acquisition: the calibration marker M is randomly moved *k* times within the working volume. **b** Reference system alignment and scaling: a calibrated artefact consisting of 3 markers (M_1, M_2, and M_3) is used as reference target. ($u_{i,j}$, $v_{i,j}$) are the 2D coordinates of the projection of the *j*th marker ($j = 1,\ldots,3$) on the *i*th camera ($i = 1,\ldots,4$). **c** Calibration output: camera positions C_i (indicated by *squares*) and orientations (indicated by *bold dashed lines*) are provided

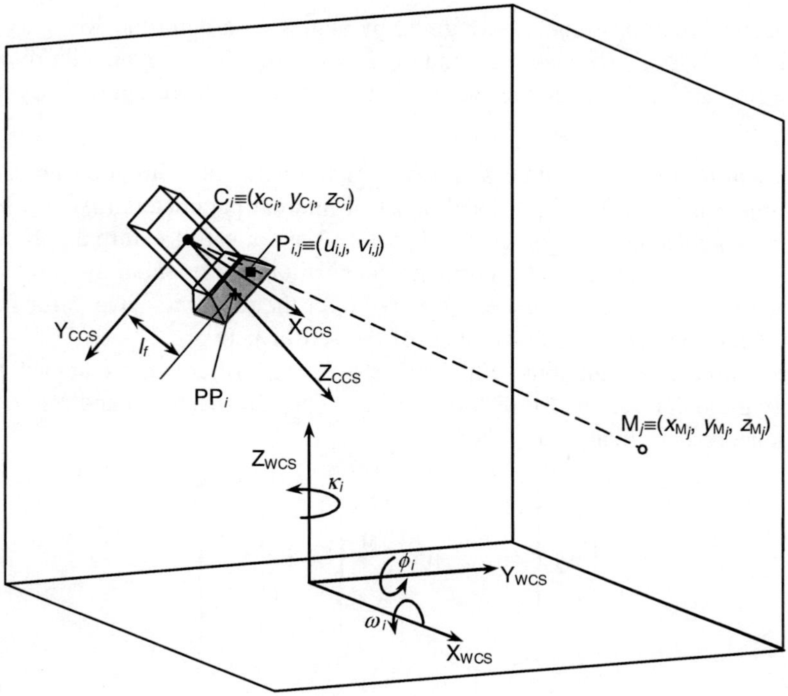

Fig. 3.20 Setup of a generic, camera-based, localization problem in the 3D space. The camera projection centre $C_i \equiv (x_{Ci}, y_{Ci}, z_{Ci})$ and the 3D marker $M_j \equiv (x_{Mj}, y_{Mj}, z_{Mj})$ are defined with respect to a world coordinate reference system, having *axis lines* X_{WCS}, Y_{WCS}, and Z_{WCS}. The plus (+) and the square (■) on the camera image plane (grey area) identify the principal point PP_i and the image point $P_{i,j}$, respectively. Their coordinates (u_{0i}, v_{0i}) and $(u_{i,j}, v_{i,j})$, are defined in an image coordinate reference system F_{ICS}, having the origin in the upper left corner of the image plane. The terms ω_i, ϕ_i, and κ_i represent the three sequential rotations that align the world coordinate reference system to the local (camera-fixed) coordinate reference system, having origin in the camera projection centre C_i, axis Z_{CCS} oriented along the positive direction of the optical axis, and axes X_{CCS} and Y_{CCS} to define a plane parallel to the image plane

The ratio c_{vi}/c_{ui}, denoted as aspect ratio, allows to handle non-square pixels (if $c_{vi}/c_{ui} \neq 1$);

- $\delta u_{i,j}$ and $\delta v_{i,j}$ are the total lens distortions, which allow to correct measured image coordinates to take into account deviations from the ideal perspective camera model due to imaging errors. Correction models are generally applied to the measured image coordinates, taking into account lens distortion effects. They generally consider both radial and tangential distortion effects through polynomial series, defined by a vector of distortion coefficients $\boldsymbol{k}_{Ci}$ (Brown 1971; Heikkilä and Silvén 1997);
- $r_{k,l,i}$ (with $k, l = 1,\ldots,3$) are the elements of the rotation matrix $\boldsymbol{R}_i \in \mathbb{R}^{3,3}$, which relates the world coordinate reference frame F_{WCS} and a local coordinate reference frame F_{CCS}, having origin in the camera projection centre and axes X_{CCS}, Y_{CCS}, and Z_{CCS}. This matrix can be obtained by sequentially applying three

elementary rotations, commonly given by angles ω_i around the X_{WCS} axis, ϕ_i around the Y_{WCS} axis, and κ_i around the Z_{WCS} axis. The set of angular rotations $\Omega_i = [\omega_i, \phi_i, \kappa_i]^T$ thus defines the orientation of the camera optical axis in the 3D space.

It is worthy to observe that, as a result of the network calibration procedure, each camera is characterized through a set of internal parameters (u_{0i}, v_{0i}, l_{fi}, c_{ui}, c_{vi}, $\boldsymbol{k}_{Ci}$), which represents its technical features, and a set of external parameters (x_{Ci}, y_{Ci}, z_{Ci}, ω_i, ϕ_i, κ_i), which represents its position and orientation with respect to the world coordinate reference frame. We refer the reader to Chap. 5 for further details about the calibration procedure of the sensor network.

The collinearity equations (Eq. 3.11) can be re-written in a compact form, neglecting the corrections for distortion effects ($\delta u_{i,j} = \delta v_{i,j} = 0$) and referring to homogeneous coordinates, as follows:

$$\begin{bmatrix} u_{i,j} \\ v_{i,j} \\ 1 \end{bmatrix} = \mu_i \boldsymbol{K}_i \boldsymbol{W}_i \begin{bmatrix} x_{M_j} \\ y_{M_j} \\ z_{M_j} \\ 1 \end{bmatrix} = \boldsymbol{P}_i \begin{bmatrix} x_{M_j} \\ y_{M_j} \\ z_{M_j} \\ 1 \end{bmatrix} \tag{3.12}$$

where:

- μ_i is a non-zero scale factor;
- $\boldsymbol{K}_i \in \mathbb{R}^{3,4}$ is the matrix of internal parameters, defined as:

$$\boldsymbol{K}_i = \begin{bmatrix} l_{f_i} c_{u_i} & 0 & u_{0_i} & 0 \\ 0 & l_{f_i} c_{v_i} & v_{0_i} & 0 \\ 0 & 0 & 1 & 0 \end{bmatrix} \tag{3.13}$$

- $\boldsymbol{W}_i \in \mathbb{R}^{4,4}$ is the matrix of external parameters, defined as:

$$\boldsymbol{W}_i = \begin{bmatrix} & \boldsymbol{R}_i & & \begin{matrix} -x'_{C_i} \\ -y'_{C_i} \\ -z'_{C_i} \end{matrix} \\ 0 & 0 & 0 & 1 \end{bmatrix} \tag{3.14}$$

being (x'_{Ci}, y'_{Ci}, z'_{Ci}) the coordinates of the projection centre C_i in the local camera reference frame F_{CCS}.

- $\boldsymbol{P}_i \in \mathbb{R}^{3,4}$ is the so-called camera projection matrix.

The localization problem consists in determining the 3D coordinates of a marker M_j according to the image views of different cameras. According to Eq. 3.11, if one only camera view is available, only the direction to the point M_j can be determined. Therefore, a localization algorithm entails observing the same

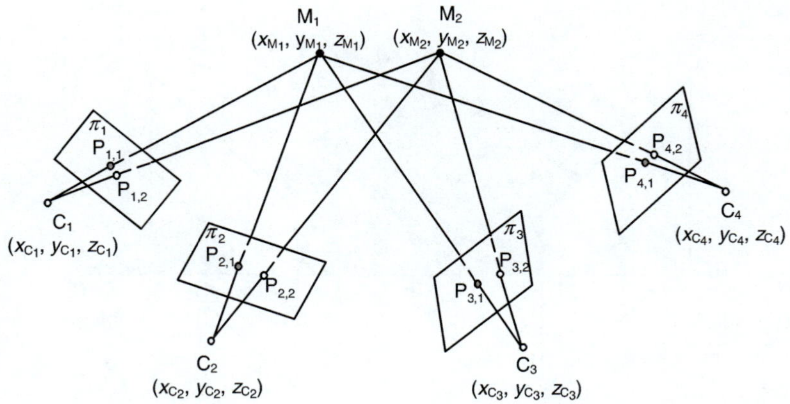

Fig. 3.21 Graphical representation of the localization problem when a setup of four cameras ($n_c = 4$) is used to reconstruct the 3D position of two markers ($m = 2$). C_i (with $i = 1,\ldots,4$) and M_j (with $j = 1,2$) refer to the 3D coordinates of the *i*th camera projection centre and the *j*th marker, respectively. Point $P_{i,j}$ represents the 2D projection of M_j on the projection plane π_i of the *i*th camera. It corresponds to the intersection of the camera plane π_i with the projection line of M_j (i.e., the line passing through the 3D point and the camera projection centre) (adapted from Galetto et al. 2011). (with permission)

point from at least two different camera positions. Marker coordinates can thus be computed as the intersection of the two spatial directions.

More generally, given a calibrated camera layout (i.e., n_c cameras, with known internal and external parameters) focused on m markers, for each m-uple of 2D pixel coordinates $P_{i,j} \equiv (u_{i,j}, v_{i,j})$, with $i = 1,\ldots,n_c$ and $j = 1,\ldots,m$, the localization algorithm has to provide the 3D coordinates of the corresponding m retro-reflective markers (Fig. 3.21). The localization procedure, which follows from the fundamentals of digital photogrammetry (Mikhail et al. 2001), is articulated in three main steps:

1. find the correspondences among pixels in different image views;
2. match the 2D information of different camera views for recovering the spatial coordinates of probe targets;
3. reconstruct the 3D position of the probe tip.

Step 1: Finding Point Correspondences

The first step consists in reconstructing the matrix $\boldsymbol{H}$ of 2D pixel coordinates corresponding to the projection of the same 3D point onto the image planes of different cameras. Since the generic marker M, positioned at unknown 3D coordinates, might not be visible from all cameras (e.g., because out of the field-of-sensing or shadowed), $\boldsymbol{H} \in \mathbb{R}^{p,2}$ where $2 \leq p \leq n_c$.

Epipolar geometry, i.e., the intrinsic projective geometry between two views, has been used to correlate information from multiple camera images (Longuet-Higgins 1981; Hartley and Zisserman 2004). Referring to Fig. 3.22, the correlation between two 2D pixels, $P_1 \equiv (u_1, v_1)$ and $P_2 \equiv (u_2, v_2)$, detected by

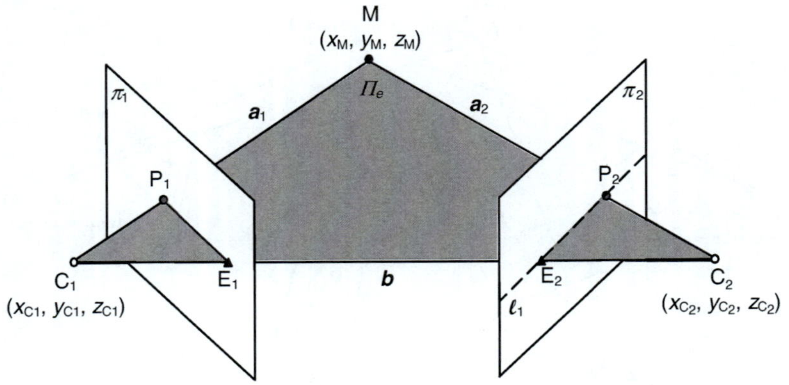

Fig. 3.22 Epipolar geometry principles. C_1 and C_2 (*white circles*) are the camera projection centres, M (*black circle*) is the 3D point. P_1 and P_2 (*grey circles*) represent the 2D projections of M on the projection planes (π_1 and π_2) of camera C_1 and C_2, respectively. The line joining each camera centre is named base vector $\boldsymbol{b}$. Points E_1 and E_2 (*black triangles*) are the epipoles of the two cameras, i.e., the intersection of the base vector with the camera projection plane. The plane passing through the camera centres, the 3D point, and the 2D image points is denoted as the epipolar plane (Π_e). The intersection of Π_e with the projection plane of a camera is the epipolar line. For example, the *dashed line* $\boldsymbol{l}_1$ is the epipolar line related to point P_1 on the projection plane of camera C_2 and represents the set of possible locations of the corresponding point P_2 (adapted from Hartley and Zisserman 2004). (with permission)

two different cameras (denoted as C_1 and C_2), states to what extent they can be considered as the projections of the same 3D point M onto the camera planes.

According to epipolar geometry principles, the camera centres C_1 and C_2, the 3D point M and the corresponding 2D pixels P_1 and P_2 in the camera projection planes π_1 and π_2, are coplanar. The coplanarity condition can be expressed through an equation that forces these points to lie on a common plane (the epipolar plane Π_e). Being $\boldsymbol{a}_1 = [(x_M - x_{C1})\ (y_M - y_{C1})\ (z_M - z_{C1})]^T$ and $\boldsymbol{a}_2 = [(x_M - x_{C2})\ (y_M - y_{C2})\ (z_M - z_{C2})]^T$ the vectors connecting the centres of camera C_1 and C_2 to the 3D marker M, and $\boldsymbol{b} = [(x_{C2} - x_{C1})\ (y_{C2} - y_{C1})\ (z_{C2} - z_{C1})]^T$ the vector connecting the two camera centres (base vector), the coplanarity condition can be formulated as a triple scalar product:

$$\boldsymbol{a}_1 \cdot (\boldsymbol{b} \times \boldsymbol{a}_2) = \boldsymbol{a}_1^T \boldsymbol{T} \boldsymbol{a}_2 = 0 \tag{3.15}$$

where the term $\boldsymbol{T}$ represents the skew-symmetric matrix referring to the base vector $\boldsymbol{b}$:

$$\boldsymbol{T} = \begin{bmatrix} 0 & -(z_{C_2} - z_{C_1}) & (y_{C_2} - y_{C_1}) \\ (z_{C_2} - z_{C_1}) & 0 & -(x_{C_2} - x_{C_1}) \\ -(y_{C_2} - y_{C_1}) & (x_{C_2} - x_{C_1}) & 0 \end{bmatrix} \tag{3.16}$$

According to Eq. 3.12, the distance vector $\boldsymbol{a}_i$ between the marker M and the projection centre of the generic camera C_i can be written as:

$$\boldsymbol{a}_i = \begin{bmatrix} x_{\mathrm{M}} - x_{\mathrm{C}_i} \\ y_{\mathrm{M}} - y_{\mathrm{C}_i} \\ z_{\mathrm{M}} - z_{\mathrm{C}_i} \end{bmatrix} = \boldsymbol{R}_i^T \boldsymbol{K}_i^{-1} \begin{bmatrix} u_i \\ v_i \\ 1 \end{bmatrix} \tag{3.17}$$

Applying Eq. 3.17 to explicit the terms in Eq. 3.15, the coplanarity condition can then be formulated as a matrix product:

$$[u_1 \quad v_1 \quad 1]\left(\boldsymbol{K}_1^{-1}\right)^T \boldsymbol{R}_1 \boldsymbol{T} \boldsymbol{R}_2^T \boldsymbol{K}_2^{-1} \begin{bmatrix} u_2 \\ v_2 \\ 1 \end{bmatrix} = [u_1 \; v_1 \; 1]\boldsymbol{F} \begin{bmatrix} u_2 \\ v_2 \\ 1 \end{bmatrix} = 0 \tag{3.18}$$

where the fundamental matrix $\boldsymbol{F}$ is a function of the internal parameters ($\boldsymbol{K}_1$, $\boldsymbol{K}_2$) and the external parameters ($\boldsymbol{R}_1$, $\boldsymbol{R}_2$, $\boldsymbol{T}$) of the given pair of cameras, C_1 and C_2.

It is noteworthy that, due to sensor and environmental noise sources and image discretisation, perfect coplanarity might not be verified. Therefore, the algorithm that searches for corresponding points evaluates the non-zero left-hand term d_e of Eq. 3.18:

$$d_e = [u_1 \quad v_1 \quad 1]\boldsymbol{F} \begin{bmatrix} u_2 \\ v_2 \\ 1 \end{bmatrix} \neq 0 \tag{3.19}$$

Defining as epipolar line the intersection of the epipolar plane with the image plane of each camera (see for instance $\boldsymbol{l}_1$ in Fig. 3.22), the coplanarity condition implies that the point P_2 in the image plane of camera C_2, corresponding to point P_1 on the projection plane of camera C_1, will lie on the epipolar line $\boldsymbol{l}_1$. According to this principle, the term d_e represents the distance of point P_2 from the epipolar line of point P_1. Since large distances lead to pixel correlation mismatches and large reprojection errors, a threshold method based on this distance has been implemented to find correspondences between different image views. According to this, the point correspondence between the 2D pixels P_1 and P_2 is verified if the distance d_e is less than a user-defined dimensionless threshold ε (see Fig. 3.23).

Furthermore, possible concurrent presence of more than one retro-reflective marker within the working volume could give rise to some ambiguities in measurement point recovery. In some practical cases, probe positioning with respect to the IR sensor and its orientation could correspond to a very small distance between the two pixels in an image view. In order to reduce the errors in pixel correlation, a minimum search approach has been implemented.

Following this approach, when two pixels P_2 and P'_2 in the projection plane π_2 of camera C_2 verify the threshold constraint, the point P_1 in the projection plane π_1 of camera C_1 will be correlated to the one which minimizes the distance d_e (see Fig. 3.23).

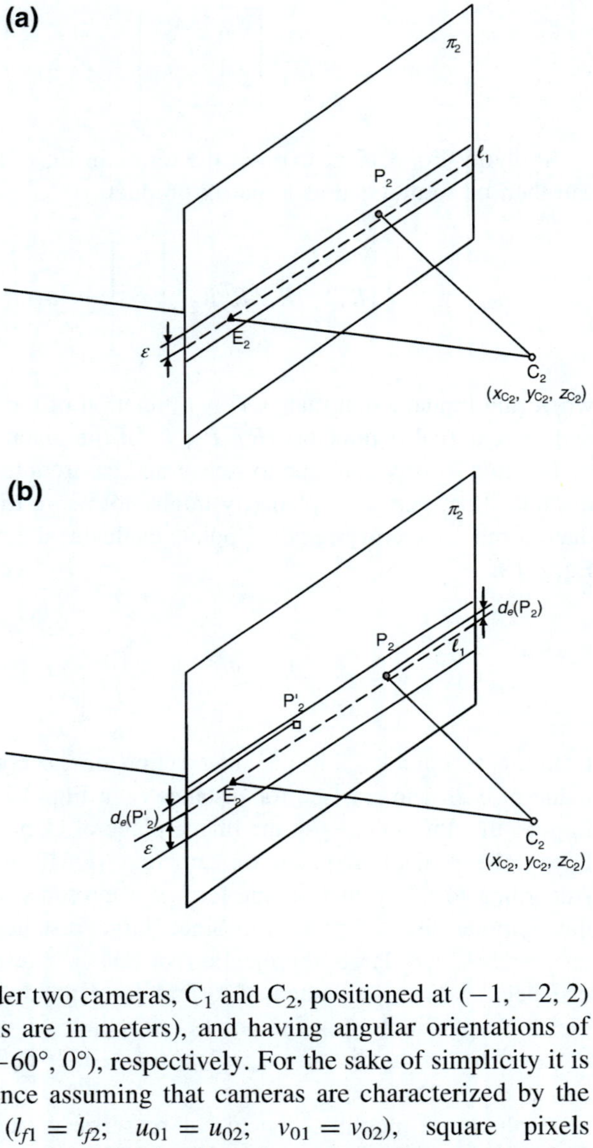

Fig. 3.23 Point correspondence issues in epipolar geometry. **a** Graphical representation of the threshold constraint. The point P_2 on the projection plane π_2 of camera C_2 corresponds to a point P_1 on the projection plane π_1 of a camera C_1 if its distance from the epipolar line $\boldsymbol{l}_1$ related to point P_1 is less than a threshold ε. **b** Graphical representation of the minimum search approach. Among all the image points on the projection plane π_2 which satisfy the threshold constraint (here denoted with P_2 and P'_2), the point P_1 on the projection plane π_1 of the camera C_1 is correlated to the one which has the minimum distance from its epipolar line $\boldsymbol{l}_1$

For example, let us consider two cameras, C_1 and C_2, positioned at (−1, −2, 2) and (1, −1, 1.5) (dimensions are in meters), and having angular orientations of (−30°, 45°, 0°) and (−30°, −60°, 0°), respectively. For the sake of simplicity it is considered $\boldsymbol{K}_1 = \boldsymbol{K}_2 = \boldsymbol{I}$, hence assuming that cameras are characterized by the same technical features ($l_{f1} = l_{f2}$; $u_{01} = u_{02}$; $v_{01} = v_{02}$), square pixels ($c_{u1} = c_{u2} = c_{v1} = c_{v2} = 0$) and negligible distortions. According to the plots shown in Fig. 3.24, one point $P_1 \equiv (-0.544, -0.077)$ and two points $P_2 \equiv (0.645, -0.116)$ and $P'_2 \equiv (0.859, -0.248)$ are visible on the image planes of camera C_1 and camera C_2, respectively. In order to search for corresponding points, the distance d_e is evaluated according to Eq. 3.19 for each pair of 2D points ($P_1 - P_2$ and $P_1 - P'_2$).

According to its formulation (see Eq. 3.18), the fundamental matrix $\boldsymbol{F}$ relating the two cameras is:

$$\boldsymbol{F} = \begin{bmatrix} 0.483 & -1.772 & -1.173 \\ -1.132 & 0.623 & -1.785 \\ -0.504 & 0.788 & -0.140 \end{bmatrix} \quad (3.20)$$

The evaluation of distance d_e according to Eq. 3.19 provides the following results:

$$d_e(P_1 - P_2) = -3.45 \cdot 10^{-5} < d_e(P_1 - P_2') = -0.371 \quad (3.21)$$

Points P_1 and P_2 can thus be selected as corresponding points in the two image planes as they show the minimum residual value with respect to the coplanarity condition. As a consequence they can be used to implement triangulation-based localization algorithms for reconstructing the 3D position of markers.

Step 2: Reconstructing 3D Position of Probe Targets

The second step of the localization algorithm deals with the triangulation problem (Hartley and Sturm 1997). Given its 2D positions in n different image planes (with $2 \leq n \leq n_c$), the 3D coordinates of a point M can be obtained by intersecting the camera projection lines (triangulation). Hence the set of $2 \times n$ equations with unknown variables $\boldsymbol{x}_{\mathrm{M}} = [x_{\mathrm{M}}\ y_{\mathrm{M}}\ z_{\mathrm{M}}]^T$ can be written as:

$$\boldsymbol{A}\boldsymbol{x}_{\mathrm{M}} - \boldsymbol{B} = 0 \quad (3.22)$$

where $\boldsymbol{A} \in \mathbb{R}^{2n,3}$ and $\boldsymbol{B} \in \mathbb{R}^{2n,1}$ are known matrices, whose elements are obtained as functions of camera parameters and the 2D pixel coordinates $\mathrm{P}_{i,j} \equiv (u_{i,j}, v_{i,j})$ (with $i = 1,\ldots,n$) of the projection of the jth marker on the projection planes of different cameras.

In practical applications, due to measurement noise and sensor hardware limits, the projection lines do not generally meet in a unique point and a least-squares minimization is needed using two or more cameras. An approximated vector of position coordinates $\mathbf{x}_{M^*}$ is thus obtained by solving Eq. 3.22.

A preliminary diagnostic function, based on the vector of residuals $v = \boldsymbol{A}\boldsymbol{x}_{\mathrm{M}^*} - \boldsymbol{B}$, has been implemented in order to evaluate the correctness of 3D positioning. Whenever they show diagnostic function values higher than a user--defined threshold, the estimated coordinates $\mathbf{x}_{M^*}, \mathbf{y}_{M^*}, \mathbf{z}_{M^*}$ of the measured point M^* are automatically discarded by the processing software (see Chap. 6 for further details on diagnostic tests).

Step 3: Reconstructing 3D Position of Probe Tip

The spatial coordinates of the probe tip V (see Figs. 3.2, 3.27) are calculated by means of Eq. 3.6, according to the geometry of the portable probe, known a priori, and the 3D reconstructed position of targets A and B.

It should be noted that, as they are based on 2D image views of different cameras, the triangulation results are affected by camera synchronization issues. The 3D point reconstruction algorithm should use the 2D position coordinates of the same point as seen by the different camera sensors at the same instant (synchronized camera sampling). However, by using the communication links connecting the IR sensors to

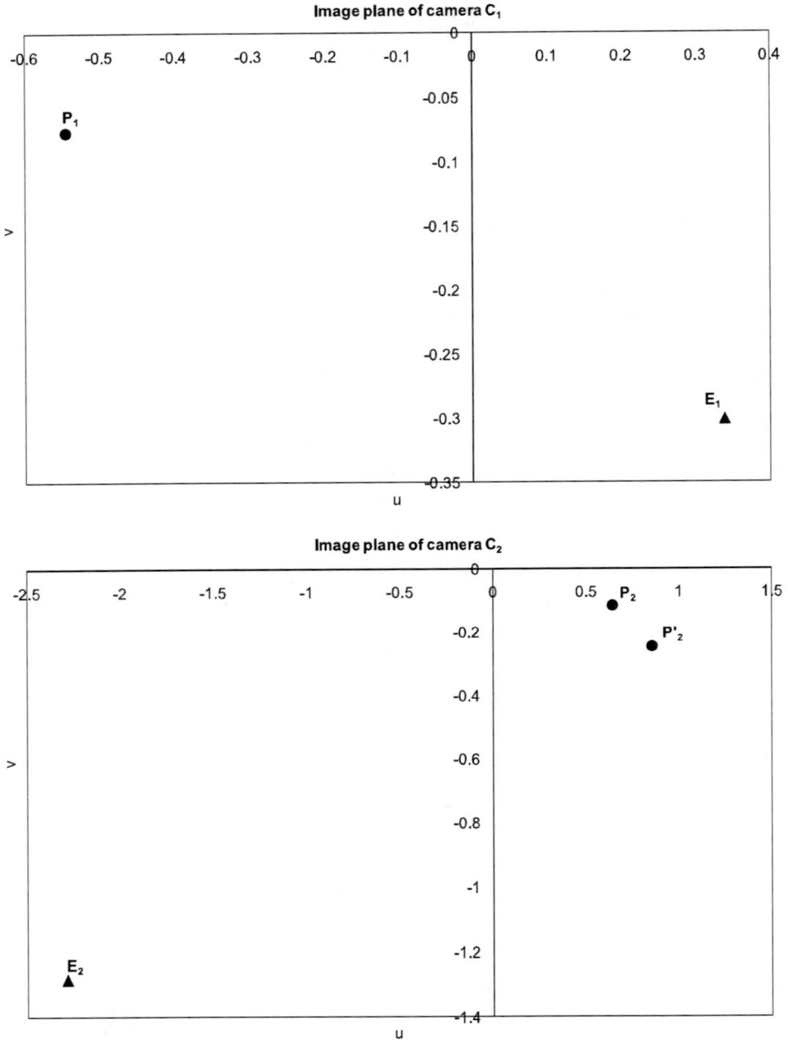

Fig. 3.24 2D representation of the image planes of camera C_1 and C_2. The triangles identify the epipoles. The *dark circles* represent the positions of the 2D pixels corresponding to the marker projections in the camera images. It should be noted that two pixels (P_2 and P'_2) are visible in the image plane of camera C_2, whereas camera C_1 is able to see just one marker, corresponding to point P_1 in its view plane

the data processing unit, a sequential sampling procedure could be implemented. Asynchronous sampling could represent a critical issue for 3D reconstruction. The higher the number of sensors the higher the total acquisition delay and thus higher possible discrepancies among different image views. Although it could represent a problem for tracking dynamic objects, sequential sampling has reduced effects on dimensional measurement performance in static conditions.

3.3.4 Prototype Development and Performance Evaluation

3.3.4.1 Prototype Description

A prototype of the MScMS-II was set up and tested at the Industrial Metrology and Quality Engineering Laboratory of Politecnico di Torino-DISPEA (Galetto et al. 2011). Low-cost IR cameras, characterized by an interpolated resolution of 1,024 × 768 pixels (native resolution is 128 × 96 pixels), a maximum sample rate of 100 Hz, and an angular FOV of approximately 45° × 30°, were chosen as sensor network devices. They are provided with an embedded tracking engine, able to perform image processing and to calculate the 2D coordinates of up to four markers.

In order to work with passive markers, each camera was coupled with a near-IR light source (Fig. 3.25), consisting of a 160-chip LED array with a peak wavelength of 940 nm and a viewing half-angle of approximately 80°. The overall sensor set (camera and LED array) weighs about 500 g and is 13 × 13 × 15 cm in size.

The IR sensors configuration can be set according to the shape and size of the measured object as well as that of the working environment (Galetto and Pralio 2010), as detailed in Chap. 4. Since marker dimensions, camera resolution, IR light source power and working volume are strictly related parameters, the sensitivity of the IR sensor set was experimentally evaluated by testing the visibility distance of differently sized retro-reflective spheres (see Fig. 3.18). Referring to the IR technology used, the system prototype demonstrated that it could track a 16 mm diameter marker in a range between $d_{min} = 50$ mm and $d_{max} = 3{,}500$ mm. On the other hand, by using a 40 mm diameter marker the traceability ranges were from 300 to 6,000 mm. Whereas the upper bound (d_{max}) of this range represents a limitation in terms of marker visibility in the camera projection plane, the lower bound (d_{min}) represents the distance under which the tracking engine was unable to correctly find the centre of the point projection in its view plane.

Given a fixed number of cameras, where all operating conditions are unchanged, the actual working volume, meaning the region within which the spatial position of a single marker can be reconstructed, depends on the technical specifications of IR cameras (e.g., resolution and focal length) and IR light sources (e.g., LED power and wavelength) as well as on the size of the markers. It should be noted that, according to triangulation principles, this volume consists of the volume of intersection of the "field-of-sensing" of at least two cameras. A network layout consisting of six low-cost IR sensors, arranged in a 5.0 × 6.0 × 3.0 m working environment according to a grid-based configuration, resulted in an actual working volume of about 2.0 × 2.0 × 2.0 m by using 40 mm diameter markers. Figure 3.26 provides a virtual reconstruction of the working layout, set up to perform dimensional measurements of a 1.5 × 0.8 × 0.5 m reference object.

The portable probe (Fig. 3.27) consisted of a rod, equipped with two reflective markers at the extremes and a stick at one end to physically "touch" the

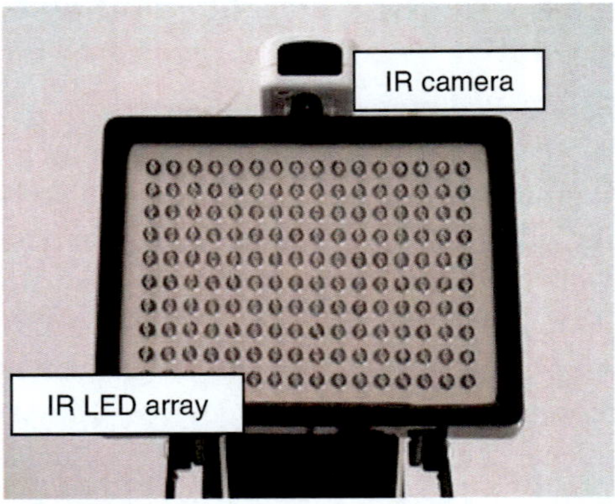

Fig. 3.25 Main components of the IR-based sensor network: an IR camera is coupled with an IR LED array to locate passive retro-reflective targets (Galetto et al. 2011). (with permission)

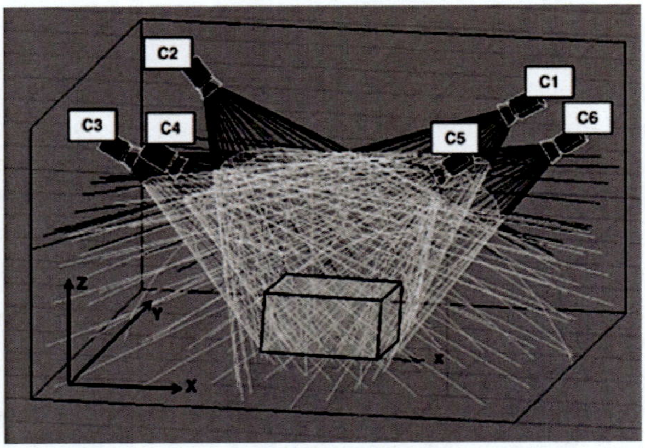

Fig. 3.26 Virtual reconstruction of the working layout. A sensor network, consisting of six IR cameras ($C_1,\ldots,C_6$), was arranged according to a grid-based configuration. The *black wireframe* represents the camera "field-of-sensing", whereas the *light grey wireframe* represents the working volume (interpreted as the volume of intersection of at least two fields-of-sensing). A 1.5 × 0.8 × 0.5 m *box*, representing a reference object to be measured, has been placed within the working environment (Galetto et al. 2011). (with permission)

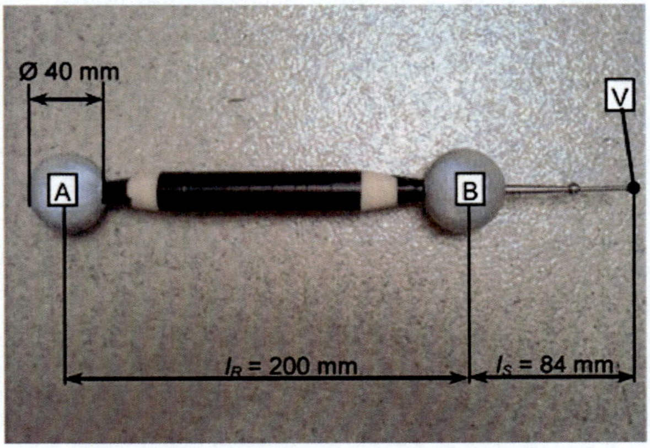

Fig. 3.27 Portable measuring probe (Galetto et al. 2011). (with permission)

Table 3.2 Standard deviation related to stability tests, with and without human operator

	$\hat{\sigma}_x$ (mm)	$\hat{\sigma}_y$ (mm)	$\hat{\sigma}_z$ (mm)
Without human operator	0.31	0.16	0.20
With human operator	0.30	0.30	0.32

measurement points. Passive markers were made by wrapping a retro-reflective silver transfer film around polystyrene spheres.

Referring to Fig. 3.27, as the probe tip (V) lies on the same line of the centres of markers A and B, the spatial coordinates of point $V \equiv (x_V, y_V, z_V)$ can be univocally determined by Eq. 3.6.

3.3.4.2 Performance Evaluation

A set of experimental tests was carried out to investigate the performance of the overall system, including the distributed sensor network, the portable measuring probe and the DPU. As they refer to a prototype implementation of the system, the results hereafter discussed could not represent a valid mean of comparison with other, industrially available, metrological solutions. In fact, they are strongly affected by the sensing hardware, the layout geometry and the network sizing. The aim of these exploratory tests was to evaluate the system performance and its dependency on the network configuration, in terms of number of IR cameras, positioning and orientation. The data herein presented has been obtained by using a set of six IR cameras, arranged in a working environment similar to the one shown in Fig. 3.26. The resulting measurement volume was about $2.0 \times 2.0 \times 2.0$ m wide. The effects of asynchronous sampling were evaluated according to a conventional dimensional measurement procedure. Considering that, even if the probe is steadied during measurement, external disturbances can occur (e.g., movements of the human operator), experimental tests were carried out to evaluate the effects of asynchronous sampling. Tests, carried out on a configuration of six non-synchronized cameras sampled at 50 Hz, showed that the acquisition delay has a negligible influence on measurement results.

As for MScMS-I the system was evaluated through stability, repeatability and reproducibility tests and characterized by a preliminary estimation of the measurement accuracy. The system stability was evaluated in five different positions, distributed all over the measurement volume, according to its definition given by the International Vocabulary of Metrology (JCGM 200:2008 2008) and reported in Chap. 1. For each position the measurement was replicated 30 times, keeping the operating conditions unchanged. The system architecture and the measurement procedure meant that human skills in handling the hand-held probe represented an external factor. In order to evaluate how it could affect system performance, stability tests were carried out with and without a human operator. Results, which are shown in Table 3.2, are reported in terms of standard deviation of the reconstructed 3D positions of the probe tip V.

Table 3.3 Standard deviation related to repeatability and reproducibility tests

	$\hat{\sigma}_x$ (mm)	$\hat{\sigma}_y$ (mm)	$\hat{\sigma}_z$ (mm)
Repeatability	0.68	0.60	0.45
Reproducibility	2.5	1.3	1.8

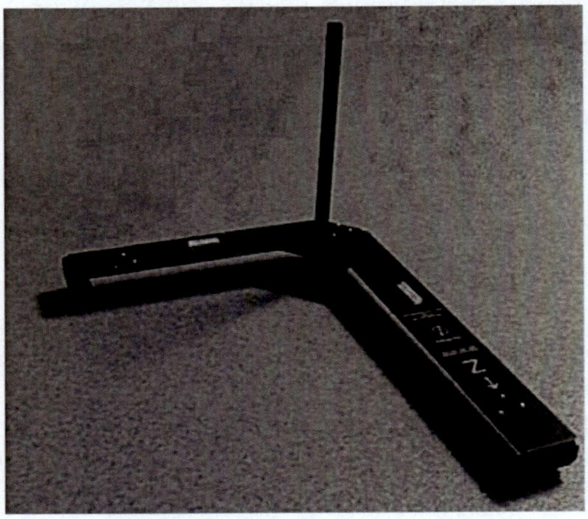

Fig. 3.28 Reference calibrated artefact for experimental evaluation of the overall system accuracy (Galetto et al. 2011). (with permission)

It is noteworthy that the human operator generally increases the variability of experimental data, depending on his/her capabilities in holding the probe in a fixed position during data acquisition. Nevertheless, as he/she actually represents a "component" of the measuring system, metrological performance was characterized including the operator in the experimental testing procedure.

Repeatability, defined in Chap. 1 according to (JCGM 200:2008 2008), was tested in 5 different points, uniformly distributed within the measurement volume. The tests were carried out by repeating the measurement 30 times for each point, repositioning the probe in the same position for each measurement. Results of repeatability tests are reported in Table 3.3, in terms of standard deviation of the reconstructed 3D positions of the probe tip V. A further source of performance degradation in terms of standard deviation is related to how skilful the human operator was at exactly replicating the probe position at each test.

Measurement reproducibility, defined in Chap. 1 according to (JCGM 200:2008 2008), was tested with reference to 5 points, distributed all over the measurement volume. For each point the measurements were repeated 30 times with different angular orientations of the portable probe. Table 3.3 reports the statistical results of these tests. As expected the standard deviation is higher for reproducibility tests than for repeatability tests. This behaviour can basically be ascribed to the influence the relative position and orientation of probe and network devices has on the overall measurement performance.

A preliminary evaluation of the overall system accuracy, defined in Chap. 1 according to (JCGM 200:2008 2008), was carried out using as reference a 3D aluminium alloy calibrated artefact (see Fig. 3.28).

On the artefact, 22 points were calibrated using a coordinate-measuring machine in order to have a set of reference points with known nominal positions. The artefact calibration and the accuracy testing were carried out by keeping the same constant environmental conditions (temperature $T = 21°C$; relative humidity $RH = 27\%$). The artefact was thus moved in five different positions, distributed within the measurement volume in order to create worst-case conditions, such as the working volume spatial limits. Taking into account the prototype hardware limitations and the severity of the experimental testing conditions, satisfactory accuracy performance was demonstrated by the MScMS-II.

The results obtained by the system prototype become particularly interesting when cost and potentiality of the metrological system are considered. Whereas its distributed architecture ensures a degree of scalability and flexibility that existing commercial systems cannot guarantee, the prototype still has significant room for improvement mainly in the area of the sensing technology. Since the state-of-the-art of IR cameras actually provide a wide choice of resolution (from less than 1 megapixel up to 16 megapixels), current CCD sensors (128×96 pixels of native resolution) could easily be replaced with higher performance models. Commercially available solutions generally enable intelligent features such as on-board 2D image analysis and processing, making the computational workload almost independent of the IR sensor resolution. Nonetheless, a trade-off between the target system performance and the economic impact of the entire system has to be found.

3.4 Comparison with Other Metrological Systems

The aim of this section is to compare the MScMS with well-tested and widespread instruments, such as classical CMMs, and with the iGPS™ (described in Chap. 2). For all these systems, measurements are taken by touching few points on the objects surface with a probe tip. Points are defined on a Cartesian coordinate system and then coordinates are processed by specific algorithms in order to determine the geometrical features of the object surface. It is recognised that, unlike CMMs, MScMS and iGPS™ are portable and easy to install, consisting of components with small dimensions distributed within the measurement volume. Furthermore, the three systems have important differences concerning technology, working principle, metrological performance and cost. Comparisons of systems are carried out according to a structured set of evaluation criteria.

3.4.1 Comparison with CMMs

CMMs are complex mechanical devices which determine the coordinates of points touched by an electromechanical probe. CMMs can be controlled either manually or by Computer Numerical Control (CNC) systems; they are available in

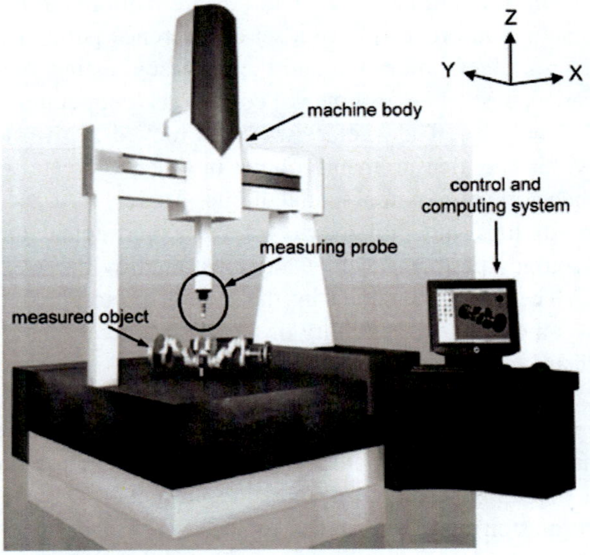

Fig. 3.29 A typical coordinate measuring machine (DEA 2010). (with permission)

a wide range of sizes and designs, offering a variety of different probe technologies. CMMs consist of three basic components (see Fig. 3.29):

- the *machine body* three carriages move the probe along the X, Y and Z Cartesian coordinate axes;
- a *measuring probe* to touch the surface points of an object to be measured;
- a *control and computing system* to calculate the Cartesian coordinates of the touched points and evaluate the shape/features of the object's surface.

CMMs are widely used in many industrial sectors to perform product control. Their reliability and accuracy has been the reason they are so widely used (Curtis and Farago 1994; Franceschini et al. 2009a). CMMs software makes it possible to perform complex types of measurement (surface construction, intersections, projections). In spite of their widespread use, these machines cannot measure every kind of object. With a few exceptions (*gantry* or *horizontal harm* CMMs, which are expensive and not portable), CMMs cannot measure large-sized objects, due to their limited measuring volume.

MScMS and CMMs will be compared according to the structured set of criteria/requirements listed in Table 3.4.

3.4.1.1 Working Volume

Working volume size. Unlike traditional CMMs, the MScMS structure is not rigidly connected. It is made of separate components that can easily be moved and arranged within the working volume according to measurement requirements. MScMS is *scalable* (or modular), since the number of network devices can be increased depending on the volume to be covered, without compromising network

Table 3.4 Comparison criteria

Working volume	Size Geometry
Setup	Portability Installation Startup Calibration and verification
Metrological performance	Dimensional measurement Other kinds of measurement
Measurement system diagnostics	On-line diagnostics Other system diagnostics
Ease of use	Automation Graphical user interface
System management	Setup phase Measuring phase
Flexibility	Kind of measurement Geometric calculation utilities Concurrent measurements
Cost	Purchasing Maintenance

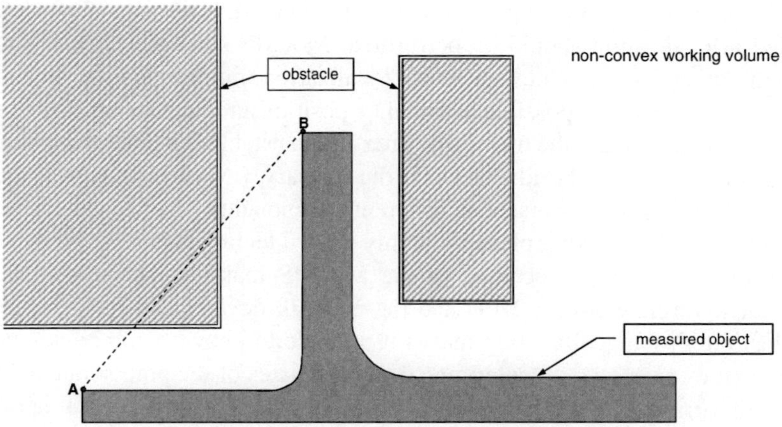

Fig. 3.30 Representation scheme of the concept of non-convex working volume (plant view) (Franceschini et al. 2009a). (with permission)

communication and slowing down measurement activities. On the contrary, CMMs are rigid and bulky systems with dimensions ranging from 1–2 to tens of meters. There is a great variety of CMMs, their working volume size can go up to hundreds of cubic meters. As discussed hereafter, performances and costs are strongly influenced by CMMs dimensions (Phillips et al. 2001).

Working volume geometry. MScMS may work in a non convex working volume, that is to say, a volume which does not contain the entire line segment joining any pair of its points (e.g., points A and B in Fig. 3.30). MScMS, due to its distributed nature, easily fits different types of indoor working environments, even with inside obstacles.

CMMs have no discontinuities in the measuring volume, since all the points within this area can be reached by the electromechanical probe. Although there are CMMs with large working volumes (i.e., *horizontal-arm* and *gantry* CMMs), the presence of obstacles in the proximity of the measured object can be problematic, since they may collide with the moving carriages. Regarding this aspect, MScMS is more flexible than CMMs.

3.4.1.2 Setup

Portability. MScMS is composed of portable lightweight devices, which can be easily installed in the area around the measured object. They can be fixed to the ceiling or mounted on standard supports and tripods (see Fig. 3.13). While the MScMS components can be moved to different operating environments, traditional CMMs are embedded in a specific working area. Once installed, CMMs have to be permanently used there. To be moved, they have to be disassembled, re-assembled, re-installed and re-started up, involving a lot of time and effort.

Installation. MScMS makes it possible to arrange network devices in different ways, depending on the application requirements. Every time the system is installed a localization should be performed. MScMS software provides a semi-automatic procedure to achieve network localization, minimizing the user's effort (see Chap. 5). It makes it possible to calculate position and orientation of the sensing devices arranged around the measuring area and to establish a Cartesian coordinate reference system (Nagpal et al. 2003). On the contrary, CMMs installation requires a great effort: the system—made of different components—has to be carried and assembled into the working place by highly skilled technicians.

Startup. The startup procedure of the MScMS makes it possible to activate communication between the DPU and the network devices and to perform probe qualification. Probe qualification makes it possible to know the probe geometrical characteristics, necessary to determine the coordinates of the points touched by the probe's tip (Franceschini et al. 2009b). This procedure is similar to that applied by CMMs at startup for activating the communication between the PC and the control system, and for selecting the probe type.

Calibration. In general, this term defines a rule that converts the value output by the instrument's sensors to values that can be related to the appropriate standard units (JCGM 200:2008 2008). Importantly, these calibrated values should be associated with corresponding uncertainties, which reliably take into account the uncertainties of all the quantities that have an influence. For MScMS, calibration is an operation that can be performed every time the system is started up. This in order to test system integrity and to set those parameters on which measurements depend (e.g., temperature and humidity). This operation does not need sophisticated instrumentation and is carried out by measuring a standard reference artefact

Table 3.5 Performance of a standard CMM in controlled *T* and *RH* conditions (DEA 2010)

Stroke x (mm)	Stroke y (mm)	Stroke z (mm)	MPE for ISO 10360/2 (μm)
500	700	500	From $1.5 + L/333$

L is the measured distance, expressed in mm. A conservative estimation of the standard deviation (σ) related to the distance measurement accuracy can be given by MPE/2 (ISO 10360 Part 2 2001). Considering the CMM in question, σ is therefore around 2 μm

with a priori known geometry. Obviously, this calibration procedure is not valid for CMMs because of the different technology and, in particular, the rigid structure. CMMs calibration cannot be accomplished directly by the user, but requires a more complex procedure defined by international standards (ISO 10360 Part 2 2001). In particular, CMMs calibration consists in a sequence of manual activities that must be carried out once or twice a year, and requires highly qualified operators and complex instruments like laser interferometers, gauge blocks or step gauges. The *network calibration* aims at establishing the initial position (and orientation if necessary) of network devices. Every time MScMS is installed, the positioning of network devices is crucial. In order to locate the portable probe, MScMS has to know the position of network devices. This step needs to be completed before performing measurements and it has strong effects on measurement uncertainty. MScMS software provides a semi-automatic procedure to achieve network calibration, minimizing the user's effort. CMMs do not need such a procedure, due to its different system architecture.

3.4.1.3 Metrological Performance

Dimensional measurement. The metrological performance of MScMS is significantly inferior to that of classical CMMs. However, it is important to notice that (1) MScMS is aimed at measuring large-sized objects and—in this context–higher measurement errors can be tolerated, (2) MScMS is a still not completely developed system at the prototype stage, and (3) the technology used by MScMS (i.e., US or IR sensors) is probably the main reason for the low accuracy of the system if compared to CMMs (Franceschini et al. 2009a). Regarding CMMs, their performance may change depending on many factors like machine dimensions, climatic conditions or probe speed of contact. Nevertheless CMMs are some order of magnitude more accurate than MScMS. To provide an example of CMMs standard performance, Table 3.5 reports the Maximum Permitted Error (MPE) on distance measurements related to a standard CMM machine (DEA 2010). In general, the MPE grows up with the size of the CMM.

Other kinds of measurements. While CMMs have been designed with the purpose of performing only dimensional measurement, MScMS can carry out other kinds of measurement. More precisely, the portable probe may be equipped with additional sensor boards. This makes it possible to associate single position measurements with other kinds of measurement, such as light intensity, temperature, acceleration, magnetic field, pressure, humidity or noise pollution (Galetto et al. 2010b). Accuracy of these kinds of measurement depends on embedded sensors performance (Crossbow Technology 2010).

3.4.1.4 Measurement System Diagnostics

On-line measurement diagnostics. MScMS, as all measurement systems, is sensitive to external factors, such as environmental conditions (temperature, humidity, presence of obstacles among distributed devices). MScMS software provides some diagnostic tools to control measurement activities and assist in the detection of abnormal functioning. Firstly, it gives the opportunity of displaying the data obtained by the network devices, making it possible to discover abnormal functioning of the system components. Secondly, it makes it possible to visualize the set of network devices able to see the probe positioning targets. This helps the operator to check whether the probe is in the optimal position to perform a specific measurement (i.e., if it communicates with at least the minimum number of network devices needed for probe localisation). Furthermore, a diagnostic tool for filtering "wrong" measurements is implemented. In the case of MScMS-I, for instance, diagnostics applies to the effects of ultrasound reflection, diffraction, or other measuring accidents on measured distances among Cricket devices (Moore et al. 2004).

On the other hand, CMMs typically offer on-line diagnostics for shape measurements: if the reconstructed shape does not reasonably fit the measured points, then a warning signal is reported. This kind of diagnostics is only possible when there is a significant measurements redundancy (for example five or more points to construct a sphere or four or more to construct a circumference). Similar diagnostic tools are implemented for MScMS.

Other measurement diagnostics. Both CMMs and MScMS can provide very similar off-line diagnostic tools. These diagnostics are based on the concept of measurement replication: if variability is higher than expected, measurements are considered not reliable (see Chap. 6). During a measurement cycle some reference distances, known a priori, are measured at regular intervals. If the variability of these points measurements is larger than expected, the measurement cycle stops, because this is a sign that system performance is deteriorating. As a consequence whenever a stop occurs, the operator has to investigate its reason. Although performed during the measurement cycle, these diagnostics cannot be considered on-line, since they require additional measurements with respect to those related to the measurand.

3.4.1.5 Ease of Use

Automation. MScMS and traditional CMMs are equipped with software packages that automate data processing. MScMS is designed to be operated purely manually: the user brings the portable probe to the object in order to touch a set of points on its surface. This is an important difference from CMMs, which are typically controlled by CNC. CMMs software makes it possible to create routines to automatically perform the same measurements on nominally "identical" objects. This implies a large reduction of time and costs when the number of (identical) objects to be measured is large. By means of a self-learning learning tool, the user can also choose to manually measure the first object allowing the system to learn the measurement patch to be repeated.

At the moment, MScMS software does not provide the same facility, since measurements are taken manually. A robot able to handle the probe and perform measurements automatically is still under development.

Graphical user interface. Both instruments (CMMs and MScMS) provide a software user interface. Their functions are based on a similar structure, with the aim of guiding the user through the various activities. Table 3.6 summarizes the results of a comparison between the MScMS and CMMs graphical user interfaces. As for CMMs, MScMS software has been developed to help operators by:

- leading them through the startup and measuring activities;
- providing tools and functions which simplify their work;
- displaying the results in a clear and complete way.

3.4.1.6 System Management

Setup. Before performing measurements, both the systems need to be setup. Regarding MScMS, the operator has the possibility of placing the network devices freely around the workpiece, taking care of their density and setting their orientation in order to adequately cover the measuring area. After this, a semi-automatic localization procedure can be performed to locate the network devices. To summarize, this procedure consists in measuring an artefact with known geometry, in different positions within the working volume. On the other hand, the setup procedure for CMMs is much more complex and requires highly skilled technicians and complex instruments (such as interferometric laser trackers).

Measuring. For both systems, the measuring phase is quite user-friendly. Regarding MScMS, the system makes it possible to modify the working volume depending on the measurement task (e.g., when the workpiece is moved or replaced with a different one), simply adding or moving some of the network devices. Of course, every time the position of one or more network devices is changed, the setup phase should be performed again. On the contrary, CMMs are rigid systems in which the working volume size is fixed.

3.4.1.7 Flexibility

Kinds of measurement. Considering flexibility as the ability of performing different types of measurement, MScMS is more flexible than classical CMMs. As

Table 3.6 Comparison between MScMS and CMMs management software applications

Operating steps	Activities	Management software tools	
		MScMS	CMMs
System startup	System initialization	Semi-automatic procedure to open the communication link	Semi-automatic procedure to start up the measuring machine
System presetting	Probe qualification	Definition of the probe's geometrical features	Semi-automatic procedure for probe qualification
	Network calibration	Semi-automatic procedure Display and storage of the network layout	–
Dimensional measurement	Choice of the measuring activity	Single shape measurement Relationships among different shapes (e.g., distances, intersections or angles)	idem
	Selection of the shape/ relationship to measure	Dedicated menu	idem
	Measurement execution	Measurement setting and execution	idem
	Audio-visual signals	Warning signals Display of the probe's communication range and the connectivity of the network	Warning signals
	Output display	Numerical and graphical display of the measured points 2D and 3D views Numerical and graphical display of the object's features Real-time measurement diagnostics	idem

described above, MScMS offers the possibility of simultaneously performing different measurements (light, acoustic noise, pressure, temperature, acceleration, magnetic field and humidity), associating them to the position measurement. These capabilities of sensor fusion, which cannot be achieved with a classical CMM, can be useful for mapping indoor environments, for example (Fischer et al. 2001; Lilienthal and Duckett 2004; Safigianni and Tsompanidou 2005).

Geometric calculation utilities. The software functions offered by MScMS are very similar to those offered by classical CMMs:

- *single shape measurement* In this case the measured workpiece's feature corresponds to a precise geometric shape (e.g., circle, plane, cylinder);
- *relationships among different shapes* The measured feature arises from a relationship between two or more different parts of the object's shape, like distances, intersections or angles between curves/surfaces.

Concurrent measurements. A significant quality of MScMS is the flexibility of its network devices. They are light, small and cheap and have an embedded processor to perform easy computations. With this distributed computational capacity, MScMS can simultaneously support two or more probes, in order to execute concurrent measurements. It is therefore possible to perform simultaneous measurements on a single object or even on different objects, improving the system sample rate. As the MScMS network is scalable and can assume different topologies, different operators can measure different objects in different parts of the network. On the contrary, CMMs are not able to simultaneously perform more than one measurement at a time.

3.4.1.8 Cost

Purchasing. Cost is a point in favour of MScMS. Its components (network devices, supports and booms, adapters) have an individual cost of the order of some tens of euros. As a consequence, the overall cost of the system is in the order of some thousands of euros. On the other hand, the cost of classical CMMs—even the most economical and simple—is one or two order of magnitude higher.

Maintenance. The MScMS system does not need very complicated maintenance. Maintenance costs are low since the system does not require the intervention of highly qualified operators. Activities of calibration and verification can easily be carried out by the user. CMM maintenance, in contrast, is a much more complicated activity, since it requires highly qualified operators and complex instruments like laser interferometers or step gauges, in accordance with ISO 10360 Part 2 (2001) standard.

Table 3.7 Comparison of technical features of MScMS and iGPS™

Technical feature	MScMS-I	MScMS-II	iGPS™
Measured variables	Distances among network devices and probe devices	Angles between sensor devices and positioning targets	Two angles among each couple of sensor and transmitter
Localisation technique	Trilateration	Triangulation	Triangulation
Sensor communication range	Up to 6 m	Up to 6–8 m(*)	More than 30 m
Number of network devices	One per every m^2	One per every 6 m^2	4 or 5 per every 400 m^2
Sample rate	About 3 points per second	Up to 50 points per second	About 50 points per second
Major sensitivity to environmental conditions	Temperature, humidity, external US sources, reflective obstacles	Light, vibrations, external IR sources, reflective obstacles	Temperature, light, vibrations
Network localisation	Semi-automated procedure	Semi-automated procedure	Semi-automated procedure
System diagnostics	Filtering of wrong measurements and to correct parameters	Filtering of wrong measurements and to correct parameters	Fixed sensors to determine whether measurement system is going out of tolerance
System calibration check	Automatic calculation of the speed of sound during measurements	Real-time adjustments of the scale	Real-time adjustments of the scale
Metrological performances	Position accuracy of about 10–20 mm (measurement of a single point by a single sampling)	Positioning accuracy of about 5 mm (measurement of a single point by averaging a number of sampling scanned in 1 s)	Position accuracy of about 0.5 mm (measurement of a single point by averaging a number of scanned in 2 s)
Working volume size	Scalable	Scalable	Scalable
Cost	Low cost	Low cost	High cost

(*) The communication range of the sensing devices implemented in MScMS-II depends on positioning target size. The reported values refer to a 40 mm diameter retro-reflective sphere

3.4.2 Performance Comparison with iGPS^{TM}

Similarly to MScMS, the system components of iGPS^{TM} are a number of transmitters, a control centre, sensors and receivers (Kang and Tesar 2004). The distributed nature of MScMS and iGPS^{TM} eases handling and provides scalability for the coverage of the measuring area. Attention should be drawn to the fact that MScMS is a still not completely developed system in prototype stage. On the other hand iGPS^{TM} has been on the market for several years.

A synthetic list of technical features of MScMS and iGPS^{TM} is given in Table 3.7. The most significant of these items are individually described in the following subsections.

3.4.2.1 Number of Network Devices

For both MScMS and iGPS^{TM}, the number of the network devices depends on their communication range and the measurement volume. In the case of MScMS-I, experimental results showed that the coverage of an indoor working volume about 4 m high is achievable using at least one network device per square meter depending on the workshop layout. Since the communication range of the network sensors of MScMS-II—and iGPS^{TM} especially—is larger, the density of nodes within the measuring volume is significantly lower.

3.4.2.2 Sample Rate

In terms of frequency of measurement acquisition, MScMS-I is very dissimilar from MScMS-II and iGPS^{TM}. This difference depends on the speed of the exchanged signals between network devices and probe devices. The speed of US signals is about 340 m/s, while laser signals are considerably faster ($\sim$300,000 km/s). Consequently, the sample rate of MScMS-I , which is about 2 points per second, is much lower than that of iGPS^{TM} and MScMS-II (about 50 points per second).

3.4.2.3 Network Calibration

MScMS and iGPS^{TM} make it possible to arrange network devices in different ways, depending on needs. Every time the systems are moved, that is, when the position of the network devices is changed, a network calibration should be performed. Obviously, this step needs to be completed before performing measurements and it has strong effects on the measurements accuracy. For this purpose, MScMS-I, MScMS-II and iGPS^{TM} provide three different semi-automated calibration procedures, all of them requiring few manual measurements.

3.4.2.4 System Calibration Check

Another activity to make MScMS suitable for the measurement is the system calibration check. It is well known that the speed of sound changes with air conditions in terms of temperature and humidity, which can exhibit both temporal and spatial variations within large working volumes. As a consequence, regarding MScMS-I, the speed of sound should be often measured and updated in calculations. What became clear from the tests is that the absolute uncertainty of iGPS^{TM} and MScMS-II is directly related to the quality of the scale bar measurement and its initial calibration. The procedure can be fully automated using two fixed positioning targets, which are tied to the extremities of a scale bar. The implementation of auto-calibration minimises downtime and corrects for environmental conditions in the measurement field, continuously and in real-time.

3.4.2.5 Metrological Performance

For the results of exploratory repeatability and reproducibility tests to evaluate the performance of MScMS and iGPS^{TM} we refer the reader to Sects. 2.4.1, 3.2.5 and 3.3.4 respectively. Due to its technology, iGPS^{TM} metrological performance is considerably better than that of MScMS. Considering these results iGPS^{TM} is approximately two orders of magnitude more precise than MScMS-I and one more than MScMS-II.

The technology employed, in particular the use of US transceivers to calculate the distances between the sensor devices, is responsible for the low accuracy of the MScMS-I compared to MScMS-II and iGPS^{TM} (Franceschini et al. 2009d; Chen et al. 2003). The ultrasound speed may change with the environmental conditions, depending on time and position. Furthermore, US signals may be diffracted and reflected by obstacles interposed between two devices. This is a negative effect for the measurement accuracy; however, it can be limited by the use of software compensation tools (see Chap. 7).

3.4.2.6 Working Volume Size

MScMS and iGPS^{TM} introduce an important difference in the typologies of measurement. The main difference from the traditional frame instruments (like CMMs) is that their structure is not rigidly connected, but consists of separate components that can be easily moved and arranged around the measuring area depending on the requirements. Therefore, these systems are *scalable* (or modular), since the number of network devices can be increased depending on the desired measurement environment. These characteristics, however, do not compromise the network communication and do not slow down activities such as network localisation and measurements.

3.4.2.7 System Diagnostics

As emerged before, MScMS-I and MScMS-II are sensitive to external factors, such as the environmental conditions of the measuring area (e.g., temperature and humidity for MScMS-I, light for MScMS-II, and presence of obstacles among distributed devices for both). Wrong distance measurements, due to ultrasound reflection, diffraction, or other measuring accidents among Cricket devices for MScMS-I, can be indirectly detected by an effective diagnostic test and then rejected (see Chap. 6). To filter wrong measurements due to external factors like light, temperature or vibrations, MScMS-II and iGPS™ also provide other types of diagnostic controls. The reliability of measurements dramatically increases by using multiple fixed sensors placed at a priori known positions. With these sensors the system can perform an automatic initial setup to continually correct the measurement field and determine whether the system is conforming to the desired tolerance (Kang and Tesar 2004).

Chapter 4
Positioning and Coverage of Distributed Devices

4.1 Introduction

Within the extensive literature concerning distributed sensors, the terms "deployment", "placement", "positioning", and "localisation" have generally been used with different meanings, according to the application fields and research communities. For the sake of clarity, herein these terms will be used according to the following definitions:

- sensor deployment: It refers to the method used to spatially distribute a set of sensors (e.g., random or deterministic);
- sensor placement/positioning: It consists in designing the sensor configuration, characterizing each sensor by its spatial coordinates and orientation angles (i.e., coordinates of the centre of the "field-of-sensing" and relative orientation);
- sensor localisation: It consists in calculating the spatial coordinates and orientation angles of a generic sensor through measured data.

The sensor placement problem has been extensively treated as the problem of positioning a (limited) number of sensors, homogeneous or not, with the aim of visiting either known or unknown data source targets in a known or unknown environment (Cassandras and Li 2005). For instance, refer to a hypothetical industrial metrology application scenario, such as measuring the spatial coordinates of a set of points (i.e., known data source targets) within a 3D measurement region with a set of homogeneous sensing devices (i.e., characterized by the same, known size and shape of the field-of-sensing) to locate these points. The sensor placement problem consists in configuring the network of distributed sensors (network nodes) through the position coordinates of the reference sensing element and the angular orientations of the field-of-sensing (Fig. 4.1).

Moreover, let's assume that the available region for positioning the sensing devices has been partitioned according to a 3D grid layout. Consequently, a discrete number of candidate points (generally higher than the number of sensors) is available for positioning the devices. The network configuration is thus designed

F. Franceschini et al., *Distributed Large-Scale Dimensional Metrology*,
DOI: 10.1007/978-0-85729-543-9_4,

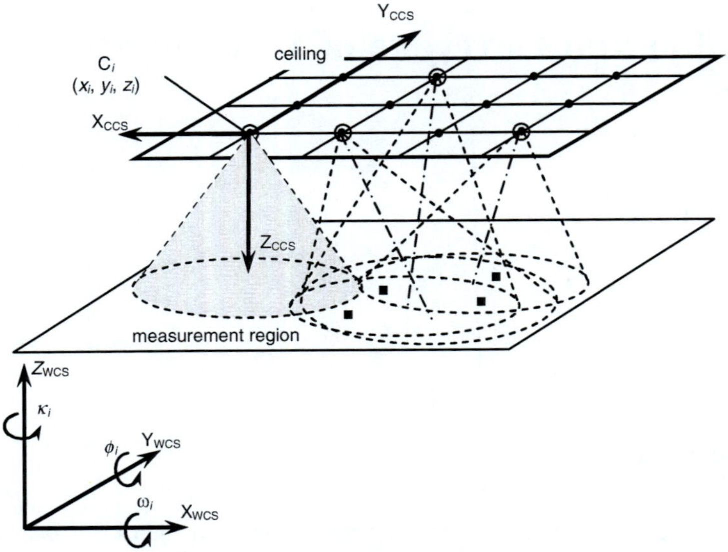

Fig. 4.1 Graphical representation of a 2D sensor placement problem in an indoor environment. The points to be measured (*filled square*) lie on a 2D area (measurement region). The region available for sensor placement is a 2D area (e.g., the ceiling of the working facility) and is partitioned according to a grid. A discrete number of candidate points for sensor placement is identified (*filled circle*), corresponding to the nodes of the grid. A possible configuration of three sensors (*open circle*) is designed, in such a way that each measurement point is within the field-of-sensing (*grey region*) of at least two sensors (localisation requirement). Each sensor is characterized by its spatial coordinates $C_i \equiv (x_i, y_i, z_i)$ in the world coordinate reference system F_{WCS}, with axis lines X_{WCS}, Y_{WCS} and Z_{WCS}. The angular orientation of its field-of-sensing is given by the set of rotations $(\omega_i, \phi_i, \kappa_i)$ that, sequentially applied to the axes of F_{WCS}, align it to a local (C_i-fixed) coordinate reference system F_{CCS} with origin in C_i, Z_{CCS} axis along the positive direction of the "sensing axis", axes X_{CCS} and Y_{CCS} to form a plane perpendicular to it

by choosing where to place sensors and how to orient them, in an attempt to satisfy the localisation requirement (i.e., minimum number of sensors needed to spatially locate a point) and possible performance metrics (e.g., measurement precision) for each point to be measured.

Different network design approaches have been applied according to the application scenario, the problem complexity, the working environment, the sensing technology, and the assigned task. Network reconfiguration capabilities and sensor deployment schemes have been adopted as main categorization criteria by Younis and Akkaya (2008).

By considering the reconfiguration features, a first classification can be made among networks consisting of mobile sensors (mobile networks), fixed sensors (stationary networks), or a combination of them (hybrid networks) (see Fig. 4.2).

Within *mobile networks*, consisting of a set of sensor nodes that may change their location with respect to an initial configuration, sensor positioning can represent a post-deployment phase or can be carried out when the network is operating. Post-deployment positioning is generally applied to improve performance of

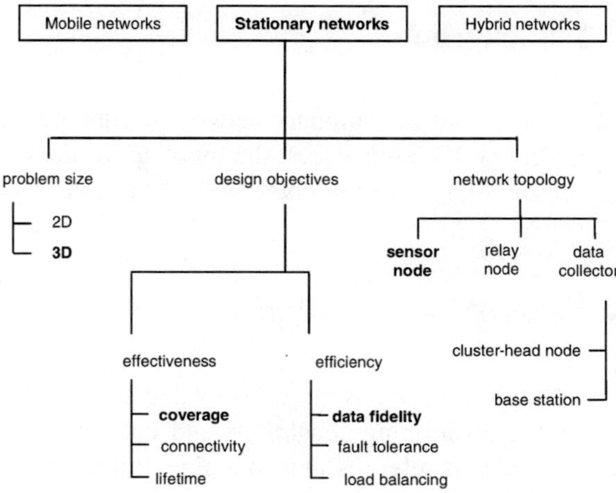

Fig. 4.2 A classification of sensor networks according to their reconfiguration capabilities. Topics in *bold* are those generally addressed in sensor positioning problems referring to distributed systems for indoor dimensional measurements (adapted from Younis and Akkaya 2008). (with permission)

randomly deployed networks (Wang et al. 2004a). On the other hand, changes of the sensor layout during network operation are commonly implemented when the environment is unknown and data sources are unknown or if they are mobile (Schwager et al. 2008). As a matter of fact, they allow for a continuous adjustment of nodes according to changes in the working environment (e.g., temperature, light, humidity), degradation of sensing performance (e.g., due to energy consumption, system failure), and changes of the mission goal (e.g., due to target mobility). In these cases, the problem consists of deploying sensors so that the information extracted from the working environment can be optimised ("coverage control" or "active sensing" problem). Different approaches to this problem have been proposed, based either on centralized control or distributed control.

On the other hand, *stationary networks*, consisting of nodes without mobility, are applied when the target locations are known and fixed. Due to their reduced flexibility, they require a thorough knowledge of the working environment and the operating conditions. The positioning strategy consists of a pre-processing phase, determining the sensor placement once before the network is operating. The network design is thus carried out according to performance metrics and sensor characteristics that are considered unchanging over time and unaffected by external factors. Unlike mobile solutions, possible changes to the application task cannot be taken into account during network operation.

Hybrid networks, involving mobile and static nodes, represent a half-way solution to take into account possible environmental changes and/or unpredictable data sources (Wang et al. 2004b).

Notwithstanding the wide variety of problem interpretations, application fields and proposed solutions, attention is hereafter focused on positioning issues related to stationary networks of sensors. Static placement is considered for distributed systems for indoor dimensional measurements of large-sized objects, as they generally have to cope with unwieldy, power consuming devices, fixed objects to be measured and pre-defined measurement tasks, in known and controllable working environments.

4.2 Background

The main issues of stationary network design are the complexity of the problem in hand (2D or 3D workspace), the topology of the reference network (heterogeneity of sensor roles), and the objectives of the sensor configuration design.

4.2.1 Problem Complexity

Problem complexity is strongly related to the geometry of the working environment, the positioning constraints and the sensor degrees of freedom. Most of the existing papers refer to 2D positioning, reducing the placement problem to a search of candidate sensor positions within an area (Yang and Scheuing 2005; Oh et al. 2007). On the other hand, a fully 3D formulation of the placement problem entails tackling with a measurement volume (instead of an area) and sensing devices having additional degrees of freedom (i.e., related to position and orientation). This challenging issue introduces further complexity to the design and could affect the applicability of efficient 2D-based algorithms (Poduri et al. 2006). Most works have addressed the 3D sensor placement problem by considering reduced-size networks (Petrushin et al. 2006), reducing the problem to a 2D formulation (Gonzales-Banos and Latombe 2001; Sinriech and Shoval 2000; Laguna et al. 2009), or referring to simple design goals (Navarro et al. 2004; Ray and Mahajan 2002).

4.2.2 Network Topology

Another issue of sensor positioning is related to the network topology. It consists in searching for an effective and efficient placement of nodes according to their role within the network. Although most of the proposed solutions aim at optimising the sensing task through networks of homogeneous nodes, a number of solutions also tackle the optimisation of heterogeneous networks. Several strategies for placing relay nodes and/or data collectors besides simple sensor nodes have been proposed (Xu et al. 2005c; Tang et al. 2006; Bogdanov et al. 2004; Pan et al. 2005). In these cases, the design strategy is strictly related to the role of the node to be placed, either it is in charge of forwarding data from one or more sensor nodes (relay node) or it has to gather, aggregate and manage data from the network (data collector).

4.2.3 Design Objectives

The main design objectives of sensor placement are related to network effectiveness (e.g., coverage, connectivity, and lifetime) and efficiency (e.g., data fidelity, fault tolerance, load balancing) (Younis and Akkaya 2008). A typical design problem can be formulated as a search procedure aimed at finding the solution which provides the best performance measure with a given number of devices. The problem of finding a layout fulfilling the design objectives can be further complicated by the requirement of minimizing the network size, i.e., the number of sensors.

Coverage, which is intended as a measure of sensing capabilities, has been the primary objective of most works on sensor placement strategies. A variety of interpretations are available in literature, depending on the field of application and the sensing technology. Coverage has been referred to continuous areas (Liu and Mohapatra 2005), discrete points (Chakrabarty et al. 2002; Dhillon et al. 2002) or crossing paths (Megerian et al. 2002; Meguerdichian et al. 2001). It can be defined as capability to detect a target ("detection coverage") or to localise it ("localisation coverage"). Performance measures of coverage have been proposed based both on coverage quantity (e.g., size of covered volumes) and coverage quality (e.g., number of sensors covering the same area/point/path). The main issues influencing coverage are the sensing model and the coverage measure.

Connectivity, which is intended as a measure of communication capabilities, is generally addressed as a key objective in order to operate with efficient sensor networks. A relationship between coverage and connectivity has been derived in (Wang et al. 2003), in which it has been also demonstrated that coverage implies connectivity whenever the communication range is more than two times the sensing range. On the other hand, several works have addressed the coverage-connectivity optimisation (Ghosh and Das 2008).

Whenever the *lifetime* objective is taken into account, the reduction of energy consumption, the sensor redundancy and the network topology become fundamental issues of the placement problem. Strategies are thus aimed at prolonging the network lifespan (Dasgupta et al. 2003; Chen et al. 2005).

Data fidelity is generally related to reduce probabilities of false alarms, entail low measurement distortion, and improve detection capabilities (Zhang and Wicker 2004; Wang et al. 2006).

In the context of distributed systems for industrial metrology applications, given an homogeneous set of sensing devices and a dimensional measurement task, the positioning problem consists in searching for a fixed network geometry to accomplish some performance metrics (e.g., network size, position accuracy, measurement precision) while fulfilling environmental and operating constraints. The network geometry is the configuration of sensors (nodes) in a 2D or fully 3D working environment, defined according to their relative positions and orientations, where needed. Network size, coverage and quality of sensing are generally addressed as design objectives. *Coverage* is the working condition under which the sensing devices are able to provide a measure of the spatial coordinates of a reference point (localisation coverage of grid points). The degree of coverage of k, i.e., the number of nodes with the same reference point within their sensing range, is used as performance measure ("k-coverage"). On the other hand, *quality of sensing* is intended as the capability to provide reliable and accurate measurements.

The measurement precision, strongly dependent on network nodes distribution in the volume, is mostly used as a measure of performance (Yarlagadda et al. 2000; Laguna et al. 2009; Galetto et al. 2011).

4.3 Sensor Positioning Issues

The network design challenges entail a multidisciplinary knowledge referring to sensing and communication models, working environment geometry, operating conditions, physical constraints, measurement procedures, measurement tasks, and localisation techniques. Within Large-Scale Dimensional Metrology applications the focus is the placement of sensors at a subset of predefined fixed sites within the working volume, so as to provide a minimum degree of coverage and satisfactory measurement precision, possibly minimizing the number of sensors. Generally network lifetime and connectivity are not primary design issues, as it is assumed that power supply can be provided through indoor infrastructures and each sensor has sufficient transmission range to reach the base station from its site. Therefore, the following analysis is limited to the effects of the sensing model, the working environment geometry, the measurement procedures and the localisation techniques.

4.3.1 Sensing Model

Whichever kind of technology is used to measure relative positioning, based on either distance or angle measurements, a precise and reliable characterization of the sensor model is required, as it affects network efficiency and measurement performance. Due to the fact that both sensing and communication units are independent components of a device, a comprehensive sensor model should encounter both of them (Akyildiz et al. 2002). Whereas the model providing a representation of the sensor detection region is strongly related to the notion of coverage, the communication model influences network connectivity, i.e., the capability of nodes to exchange data. According to Wang et al. (2003), whenever the communication range is at least twice the sensing range, the sensor model representation can be reduced to sensing abilities (Wang et al. 2003).

Sensing ability is generally directly related to distance and decreases as the target moves away from the device (Megerian et al. 2002). Notwithstanding this common approach, different levels of complexity characterize the available sensor models, according to technology, sensitivity to external factors and detection capabilities. The simplest approach ("binary disk model") models sensors as isotropic devices, i.e., whose sensing capabilities are only extended to points within a circular disk of radius *R* and not to any point beyond it (Ghosh and Das 2008) (see Fig. 4.3a). Although it represents a simplified and computationally fast

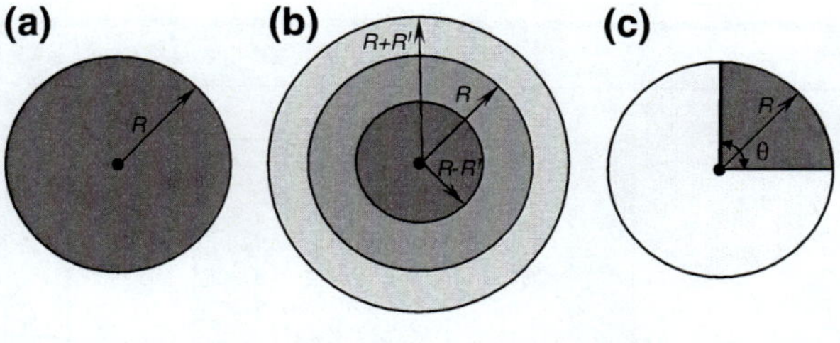

Fig. 4.3 Different models of sensor detection capabilities. **a** binary disk model: point detection is defined in a binary mode, i.e., a point is detected or not depending on whether it is within the sensing range R or not. **b** probabilistic disk model: point detection has a probabilistic distribution according to the relative distance between point and sensor device, including a measure of sensor uncertainty R'. **c** directional sensor model: the sensing area, in the two-dimensional space, reduces to a sector whose dimensions depend on detection range R and angle of sensing θ (adapted from Ghosh and Das 2008; Ai and Abouzeid 2006). (with permission)

approach, it could provide unrealistic results due to the assumption of perfect and uniform detection.

On the other hand, the "probabilistic sensing model" introduces the dependency of sensitivity on the relative target position (Dhillon et al. 2002). According to this model, the sensor capability to detect an object, R being its nominal sensing range, has a probabilistic distribution. The probability of detection p is assumed to vary exponentially with the distance between the target and the sensor (Ghosh and Das 2008):

$$p = \begin{cases} 1 & 0 \leq r \leq R - R' \\ e^{-\alpha r} & R - R' < r \leq R + R' \\ 0 & r > R + R' \end{cases} \tag{4.1}$$

where r is the distance between sensor and target, R' is a measure of detection uncertainty, and the parameter α can be used to model the quality of the sensor and the rate at which its detection probability decreases with distance (see Fig. 4.3b). As a matter of fact, perfect detection is assumed for all points that lie within a distance $(R - R')$ from the device whereas no detection is considered beyond a distance $(R + R')$.

More realistic models have been developed by taking into account detection directionality. Since a directional sensor has a finite angle of sensing θ, the sensing area reduces from the circular disk of radius R to a sector in the two-dimensional plane (see Fig. 4.3c) (Ai and Abouzeid 2006).

4.3.2 Working Environment Geometry

Within the operating environment, whose characteristics are strongly relevant for the network design, three regions can be generally identified (Laguna et al. 2009): (1) the measurement volume, (2) the sensor volume and (3) the possible unavailable volume (see Fig. 4.4).

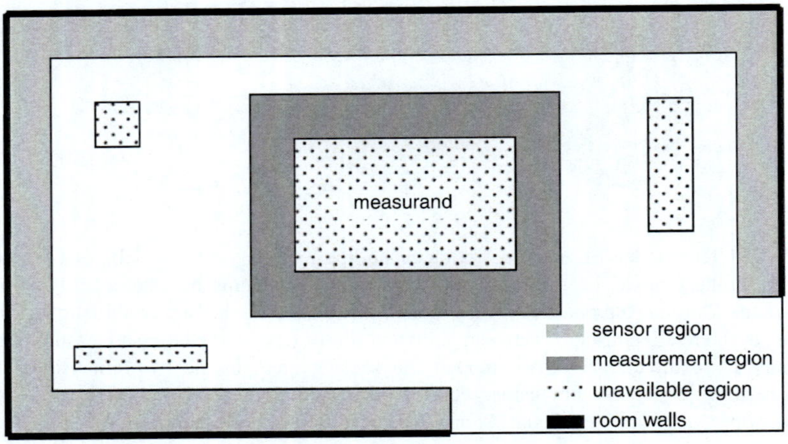

Fig. 4.4 Schematic 2D representation of the main regions characterizing a generic working environment. The *light grey area* represents the region available for sensor placement, in this case restricted to the wall proximities. The *dark grey area* identifies the measurement region, i.e., the set of measurement points. This region does not necessarily coincide with the measured object shape and generally consists of a discrete number of points. It is noteworthy that to perform measurements each point needs to be within the sensing volume of at least k sensors, being k the minimum number of sensors to perform localisation. The unavailable region (*dotted area*) corresponds to obstacles or inaccessible areas

The measurement volume represents the continuous area or the set of measurement points referring to which the sensor coverage has to be maximized. The sensor volume consists of the set of possible network device locations and could partially intersect the measurement volume. The unavailable volume is mainly represented by obstacles and physical obstructions, restraining from using this region for either placing sensors or performing measurements.

4.3.3 Aim of the Measurement

It is noteworthy that the aim of a measurement has a substantial effect on the positioning of network devices. If the goal is to reproduce the geometry of a given object, an entire volume should be adequately covered by the sensor network, whereas, if the goal is to control only some specific geometrical features, a limited number of points must be covered.

4.3.4 Localisation Techniques

As mentioned in Chap. 1, different localisation approaches can be implemented to spatially locate the sensing devices, according to the methods used for estimating the distance between two nodes (Received Signal Strength, Time of Arrival,

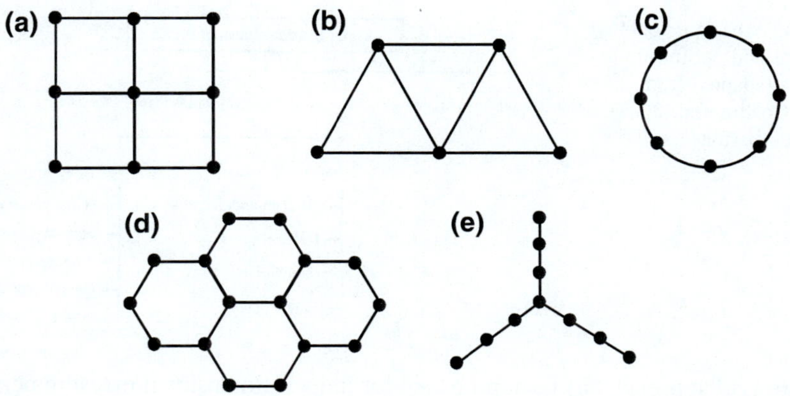

Fig. 4.5 Regular grid patterns for node placement: **a** square grid; **b** triangular grid; **c** circular grid; **d** hexagonal grid; **e** star-3 grid (Biagioni and Sasaki 2003). (with permission)

Time Difference of Arrival, Angle of Arrival) (Gibson 1999). The implemented localisation technique influences the network design through the degree of coverage requirement. As a matter of fact the minimum number of devices which have the measured point/area within their sensing varies for trilateration, triangulation and multilateration techniques (Savvides et al. 2001). As shown in Roa et al. (2007), the network performance can be strongly affected by the selected technique, as to coverage capabilities, singularities due to sensor collinearity, and precision of position estimates.

4.4 Network Design Strategies

According to the deployment scheme, random strategies, regular grid-based strategies and deterministic strategies represent the main alternatives for placing stationary sensors within a working environment (Xu et al. 2006). A random strategy for sensor placement is generally applied during recovery, detection and reconnaissance tasks, within hardly accessible, hazardous and possibly unknown environments. It generally entails very dense placement of low-cost, disposable sensors. Related research work focuses on the effects of different functions of node probability of distribution on network coverage and fault tolerance capabilities (Ishizuka and Aida 2004; Xu et al. 2005a, b).

Regular grid-based techniques have been widely applied as they are straightforward and easily scalable to large-sized problems. Different grid patterns have been proposed (see Fig. 4.5), including the equilateral triangle grid, the square grid and the hexagonal grid (Biagioni and Sasaki 2003).

In Pompili et al. (2006) the problem of achieving the maximum coverage of an underwater region with the minimum number of sensors has been faced with a triangular grid by adjusting the inter-sensor distance. In Franceschini et al. (2008)

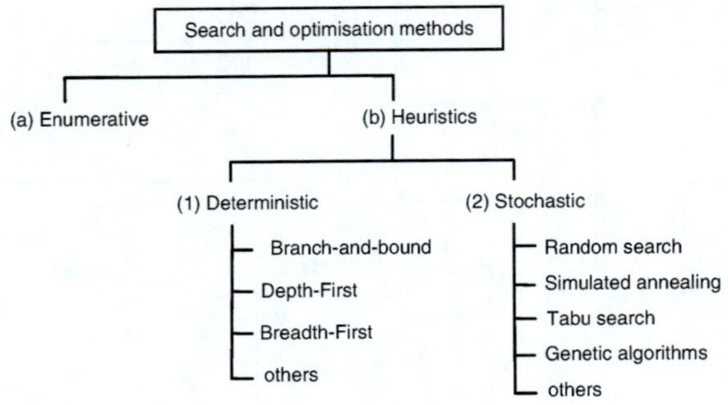

Fig. 4.6 A basic taxonomy of search and optimisation methods (adapted from Coello Coello et al. 2007). (with permission)

a square grid approach has been proposed for indoor dimensional measurements of large-sized objects through a network of US sensors. Moreover, the research in Xu et al. (2006) considers the effects of inaccurate sensor placement on the coverage performance. The minimum number of sensors to fulfil coverage requirement is derived taking into account misalignment and random errors.

On the other hand, deterministic sensor placement is usually performed in indoor, controllable and fully accessible environments. Several literature works refer to deterministic strategies applied to tasks involving acoustic, imaging and video sensor technology. Different approaches, implementing optimisation techniques, have been proposed for deterministic positioning (Padula and Kincaid 1999). Whereas regular schemes generally address the design objectives generating dense configurations, optimisation methods represent a valuable alternative when the performance measure to be optimised is not intuitive and the best trade-off among different design objectives has to be found. Moreover, they are able to take into account physical (e.g., sensing range), operational (e.g., power consumption), and geometrical (e.g., relative spatial locations between sensors) constraints. Search and optimisation methods are based either on enumerative search or heuristic search (see Fig. 4.6 for a basic taxonomy).

(a) Enumerative Search Methods

Enumerative schemes are the simplest search techniques but they are generally applied to small-sized problems, whenever all possible solutions can be characterized through a performance measure. For larger search spaces they become unfeasible due to the computational complexity of the problem (NP-hard formulation) (Mason 1997).

(b) Heuristics-Based Methods

Due to the lack of performance and the applicability limits shown by enumerative methods, heuristics are often applied to provide a consistent, although sub-optimal, solution to the sensor placement problem in "acceptable" time. Heuristic methods can be further classified as deterministic and stochastic, according to the strategy applied to limit the search space and the problem domain knowledge.

Deterministic strategies examine a set (or subset) of possible solutions at each iteration with reference to a performance metric. A "node" is expanded whenever it is considered the most promising among all the other analyzed nodes. It is quite obvious that these methods become inapplicable when they have to cope with high-dimensional multi-objective optimisation problems, involving discontinuities in the search space, multimodal objective functions, and/or NP-hard features (Coello Coello et al. 2007).

On the contrary, stochastic methods, among which random search, simulated annealing (SA), tabu search (TS) and genetic algorithms (GAs), have been widely applied for sensor placement problems. Random search is the basic strategy of such methods. Simulated annealing and tabu search are based on a neighbourhood structure and apply a selection strategy according to the performance of a solution with respect to those of its neighbors. GAs implement evolutionary principles and natural selection laws for searching among a set of possible solutions the one that best fulfils one or more performance metric(s).

(c) Random Search

Random search, which is the simplest stochastic heuristics, consists of randomly generating a number of possible solutions, evaluating their performance measures and selecting the solution which shows the best performance. This method, although characterized by a straightforward implementation and low computational complexity, is not a guarantee against multiple generations (and analysis) of the same solution. Furthermore, since one generation is independent from the other, the history of previous search steps is not taken into account.

(d) Simulated Annealing

Simulated annealing applies a search strategy based on the physical concept of "annealing" (Kirkpatrick et al. 1983; Laarhoven and Aarts 1987). Annealing consists of a heating process during which the molecules of a solid are free to randomly change their positions and a subsequent cooling process, forcing the molecules to settle to positions corresponding to lower internal stresses (i.e., lower energy state). The fitness function to be optimised is thus related to the free energy of the solid, which the search strategy aims at minimizing. Figure 4.7 shows the basic scheme of a SA implementation. The input of the search algorithm is:

- a starting point s_0, i.e., the current state of the system and initial solution of the iterative strategy;
- a starting temperature t_0;
- the cooling schedule CS, i.e., the rate at which temperature changes during the cooling process;
- the stopping criterion, which consists in defining the lowest temperature allowed.

Each possible solution of the optimisation algorithm corresponds to a system state s, to which a slight random perturbation is applied in order to change it to another state in its neighbourhood. Each state s is characterized by a fitness value $FF(s)$, which is related to one or more performance metrics.

If the new state (solution) s' shows better performance than the current one (i.e., if $FF(s') \leq FF(s)$), the move is accepted and it becomes the new reference state. Otherwise, a parameter U is randomly generated and compared to a probability $p < 1$, exponentially decreasing according to the time and the performance

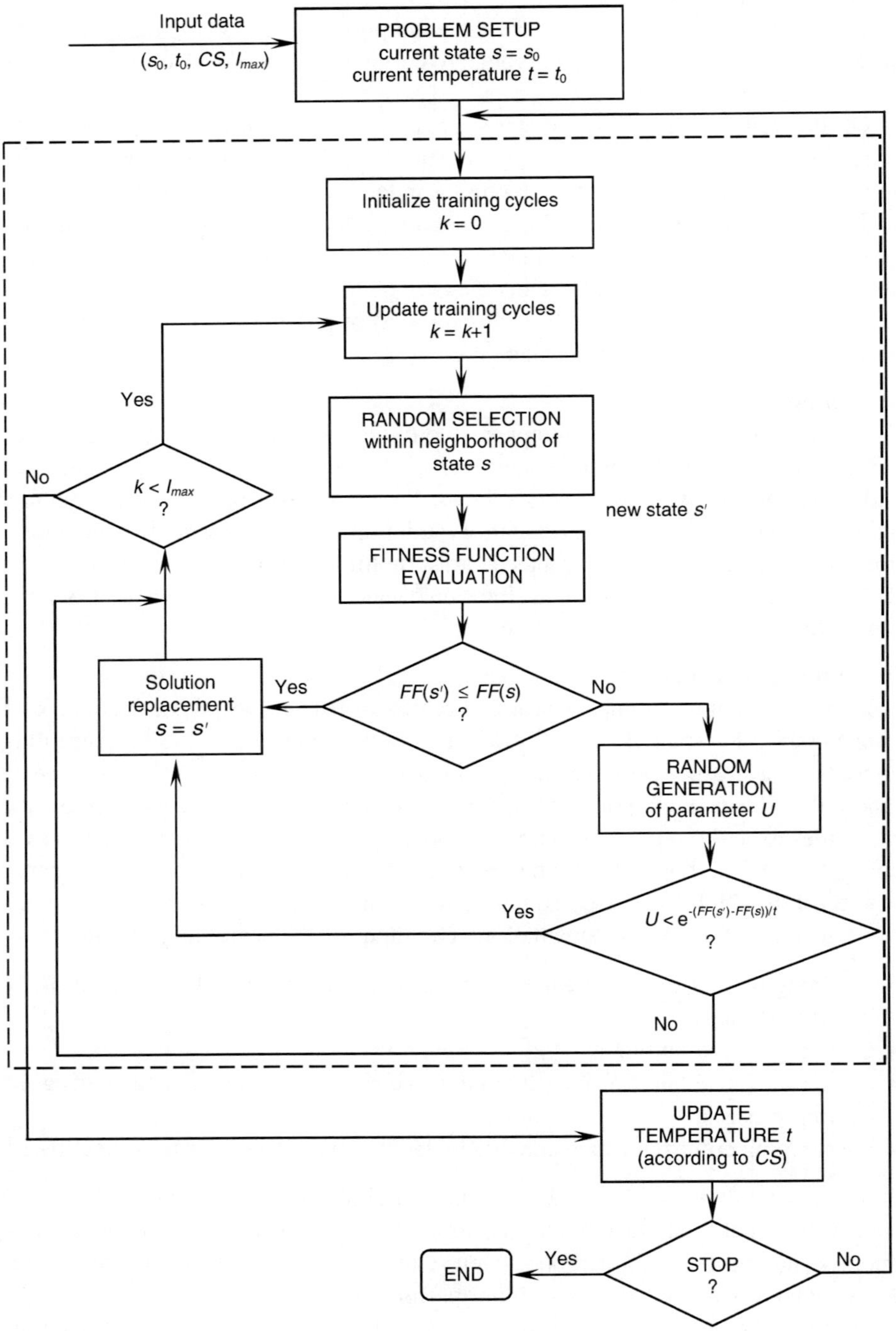

Fig. 4.7 Flow chart of a simulated annealing implementation. The starting state s_0, the initial temperature t_0, the cooling schedule CS and the number of training cycles I_{max} are the input data of the iterative procedure. $FF(s)$ represents the fitness value characterizing the state s. U is a randomly generated parameter. The *dashed box* identifies the operations of the kth training cycle

degradation compared to the previous state. If this probability is higher than the random parameter U, the undesirable move is accepted and the new solution s' becomes the new reference state anyway. Otherwise the move is ignored and the process starts again from searching in the neighbourhoods of the previous solution. Two main iterative processes are applied by this method. Firstly, for each temperature, the algorithm performs different training cycles, randomly exploring different neighbourhoods at the same time.

Secondly, once the specified number of cycles (I_{max}) have been completed, the temperature is lowered according to the cooling schedule. The simulated annealing algorithms terminates when the temperature is lower than the lowest temperature allowed (stopping criterion). It is noteworthy that the cooling schedule has a key role in the optimisation process. If the temperature changes too quickly, the algorithm could reach local minima instead of global optima. On the contrary, setting the temperature change too low involves a longer iterative procedure, making the algorithm computationally inefficient.

Different versions of the SA algorithm have been proposed to deal with multi-objective problems, focusing on the evaluation of the probability of accepting an individual according to a trade-off of its performance (Coello Coello et al. 2007).

(e) Tabu Search

Whereas simulated annealing search is partially probabilistic, tabu search is fully deterministic and allows moving to a neighbour only if it is characterized by better performance (Glover and Laguna 1997). It takes into account the search history by recording visited solutions and relative paths, optimal and near-optimal points and explored regions of the search space. This approach avoids revisiting and cycling, provides seeds for intensifying the search and supports diversification of the search space, respectively. It is noteworthy that both SA and TS are effective whenever a discretisation of the search space can be performed and best solutions tend to be neighbours, restricting the search to a subset of the design space. Hybrid approaches, e.g., combining SA and GAs or TS and GAs, have been proposed to overcome these limitations (Adler 1993; Balicki and Kitowski 2001). Generally they demonstrated an increase of computational complexity.

(f) Genetic Algorithms

Genetic algorithms are widely used whenever best solutions are expected to be scattered throughout the design space. They are computationally simple and powerful in managing large search spaces and multi-objective optimisation, without assumptions such as continuity, unimodality, existence of derivatives and convexity (Goldberg 1989; Coello Coello et al. 2007).

GAs apply mechanisms of natural selection and natural genetics to a set of possible solutions of the problem, iteratively searching for the one best fulfilling a function of one or more optimisation objective(s) (i.e., the fitness function) (Holland 1992; Goldberg 1989). Each solution (also referred to as "individual") is

(a)

chromosome 1	0	1	1	1	0	1
chromosome 2	0	0	0	1	0	0
chromosome 3	1	0	0	1	0	1
chromosome 4	1	1	0	1	1	1

(b)

chromosome 1	2.3	53.2	3.1	31.4	0.7	11.3
chromosome 2	54.3	0.2	14.3	16.5	10.2	20.9
chromosome 3	14.2	50.0	0.8	15.2	9.1	33.5
chromosome 4	18.1	17.3	30.6	8.2	16.8	7.1

(c)

chromosome 1	a	b	f	d	e	c
chromosome 2	f	d	c	b	e	a
chromosome 3	b	a	c	f	e	d
chromosome 4	d	f	a	b	c	e

Fig. 4.8 Solution encodings for an optimisation problem. *Rows* correspond to chromosomes, which represent the individuals of the population (in this example the population consists of four individuals). Columns represent the genes of each chromosome. Each gene is related to a decision variable of the problem (in this example six decision variables are considered). Some possible types of representation are reported: **a** binary encoding: the genes can only assume binary values (0/1); **b** value encoding based on scalar real numbers: each gene is a scalar entity and can assume real values; **c** permutation encoding: every chromosome is a string of entities (*numbers* or *letters*) representing a sequence

represented by "chromosomes", whose basic units ("genes") are the smallest entities the genetic algorithm is able to manage. These entities can be related to the decision variables, i.e., the parameters of the problem whose changes determine different values of the fitness function. A chromosome contains all the information needed to reconstruct the solution that it represents, as well as the related value of the fitness function ("fitness value"). This representation is also called "encoding" of the solution. Different types of representation can be implemented according to the nature of the problem, such as binary encoding (i.e., using string of bits), value encoding (e.g., using real/integer numbers, string of characters, or other complex objects), permutation encoding (i.e., string of entities representing a sequence), or tree encoding (e.g., functions or commands of a programming language). A few examples are illustrated in Fig. 4.8.

Figure 4.9 shows the basic scheme of a GA implementation.

Individuals are grouped into a population on which the genetic algorithm operates at each step of an iterative procedure ("generation"). The population size, i.e., the number of individuals per population, represents an important parameter of GA implementation as it affects its effectiveness and efficiency. As a matter of fact, whereas a too-small population provides few opportunities to search the domain for possible solutions, a too-large population dramatically increases the computational load.

The history of the search plays a key role in generating the population, as, at each iteration, a set of individuals ("parents") is selected and then combined to produce new individuals ("offspring").

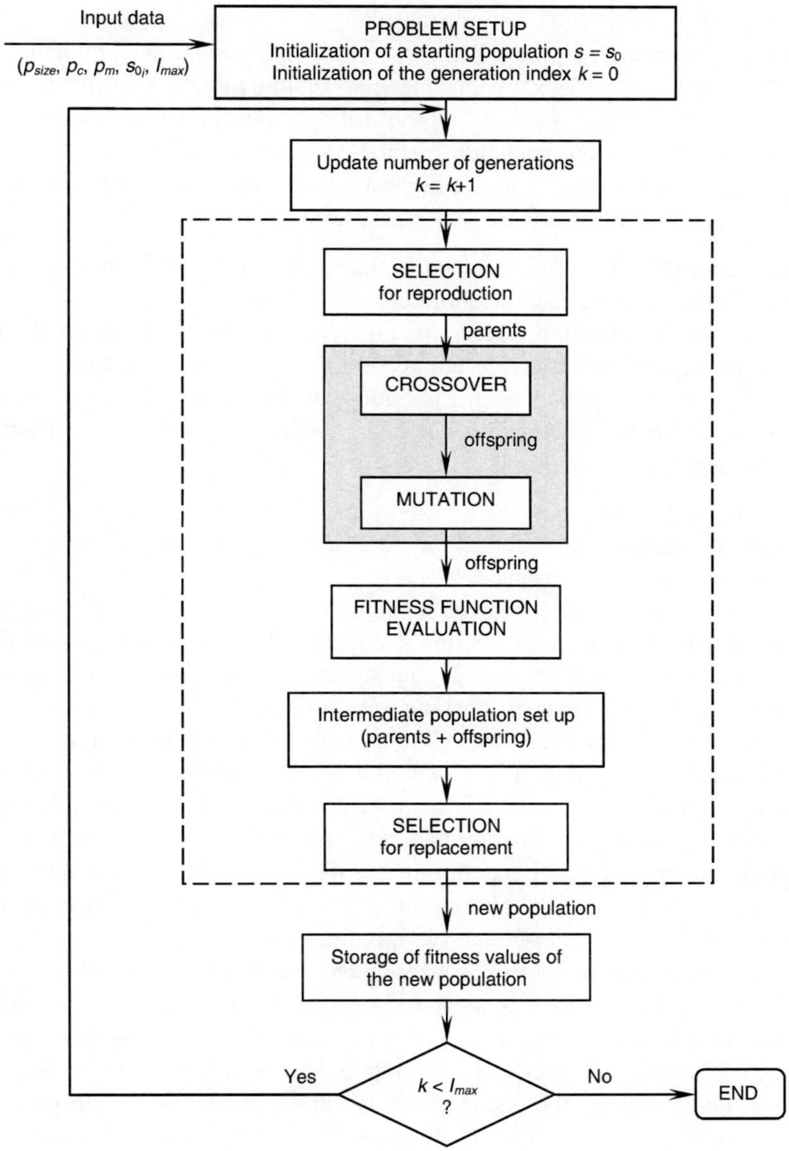

Fig. 4.9 Flow chart of a simple genetic algorithm scheme, implementing a steady state strategy, i.e., generating the new population according to a set of possible solutions including parents (individuals of past generations) and offspring (newly generated individuals). The *dashed box* identifies the genetic evolution loop, performing selection (for both reproduction and replacement), recombination and fitness function evaluation at each generation. A recombination approach based on crossover and mutation operations is considered (highlighted in the *grey box*). A starting population s_0, the population size p_{size}, a crossover probability p_c, and a mutation probability p_m are given as input of the optimisation problem. A stopping criterion based on a maximum number of iterations I_{max} is shown

Several selection methods have been proposed to choose the parents, according to either deterministic or probabilistic criteria. For instance, the "roulette wheel method" relates the probability of selecting an individual to its fitness value, whereas the "rank-based method" refers to its position in a fitness rank. On the contrary, the "tournament selection method" randomly chooses two individuals and selects as parent the one with the highest fitness value. The new population can thus be formed according to three different strategies:

- steady state: the offspring are added to the parents and then individuals for the new population are selected according to their fitness values;
- generational: the offspring replace the previous population and then individuals for the new population are selected according to their fitness values;
- elitism: a few individuals from the previous population are selected according to their fitness values and copied in the new population, which is completed with the newly generated offspring.

It is noteworthy that steady state strategy and elitism prevent the loss of the best individual of a generation, whereas the generational approach does not guarantee its survival across the iterations.

Existing individuals are combined by applying a crossover operator to generate new individuals. This operation, acting at encoding level, generally, produces two offspring from two parents by exchanging substrings. Therefore, the new solutions created through the crossover operator are somehow similar to their parents, inheriting some of their properties. A crossover probability p_c is given to define how often the crossover operator is applied. If $p_c = 0$, crossover is not applied and the offspring are exact copies of the parents. If $p_c = 1$, all the offspring are obtained by performing crossover. For a generic value $0 < p_c < 1$, a percentage of the population (p_c) is generated through crossover and the remaining (1-p_c) is simply a copy of the parents. This makes it possible to insert new features in the solutions and some individuals survive through generations.

Different crossover methods can be defined, according to the number of cross points, i.e., the number of exchanged substrings. Some examples of crossover methods are reported in Fig. 4.10. Generally referring to an m-point crossover, n being the number of genes characterizing an individual, m cross points are chosen at random along the encoded string, corresponding to the kth gene (with $k = 1,\ldots,n$). They are then sorted into ascending order avoiding duplicates.

Offspring are produced by exchanging substrings between successive crossover points. The most common crossover operator types are one-point crossover and uniform crossover (Holland 1992; Syswerda 1989). Whereas using one-point crossover two substrings from each parent are simply swapped to create two children, the uniform crossover can be interpreted as a $(n - 1)$-point crossover. As a matter of fact, for each gene of the offspring the parent from which it has to be inherited is randomly chosen with equal probability. It is noteworthy that crossover can considerably influence the performance of the optimisation algorithm and affects its computational complexity.

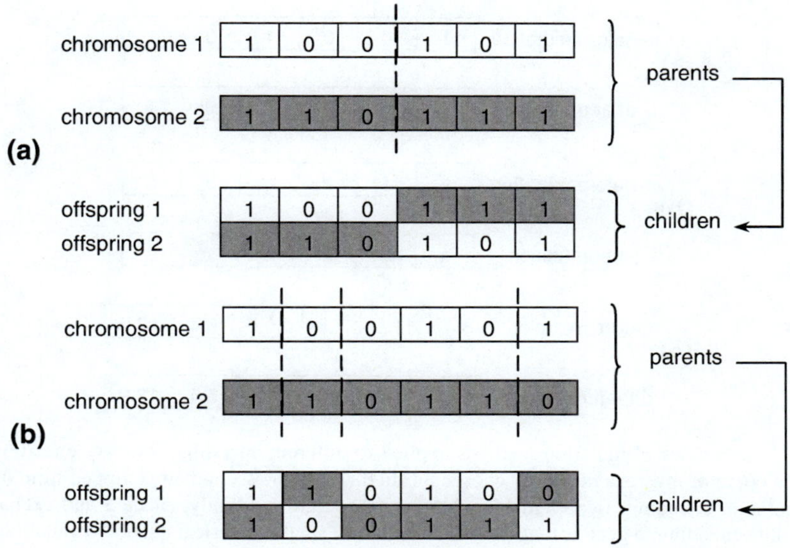

Fig. 4.10 Examples of multi-point crossover methods applied to a binary encoded solution. In a generic m-point crossover, m cross points, represented by the *dashed lines*, are randomly selected at each operation. All the substrings between two cross points are exchanged between parents to form new offspring, except the one between the first gene and the first crossover point. **a** one-point crossover ($m = 1$); **b** multi-point crossover (with $m = 3$)

Furthermore, a mutation operator can be applied to the newly generated individuals, by making small random changes to their encoding. A mutation probability p_m is given to define how often the mutation operator is applied. If $p_m = 0$, no mutation is applied and the offspring are not changed after crossover. If $p_m = 1$, all the genes are changed. For a generic value $0 < p_m < 1$, only a percentage p_m of the genes of the chromosome is mutated.

Different methods of mutating the genes of an individual are available according to the encoding used (see Fig. 4.11 for some examples). Whereas in binary encoding the mutation operator flips a randomly chosen bit from 0 to 1 or vice versa and in permutation encoding it exchanges two randomly chosen elements of the sequence, with real value encoding small quantities are randomly added to or subtracted from one or more genes. The mutation step, i.e., the numerical change to be applied to the genes, can be either constant or adapted according to the evolution history. Since it is responsible for a random perturbation of the genetic evolution, mutation plays an important role in the preservation of population diversity. It prevents convergence on local optima and is useful for an efficient exploration of the search space.

Generally, a check on generated individuals is applied at each iteration as the genetic operators could provide encoding corresponding to unfeasible solutions.

The search and optimisation problem is formulated through a fitness function, whose values are computed at each generation for each individual of the population. The fitness values are representative of the capability to address one or

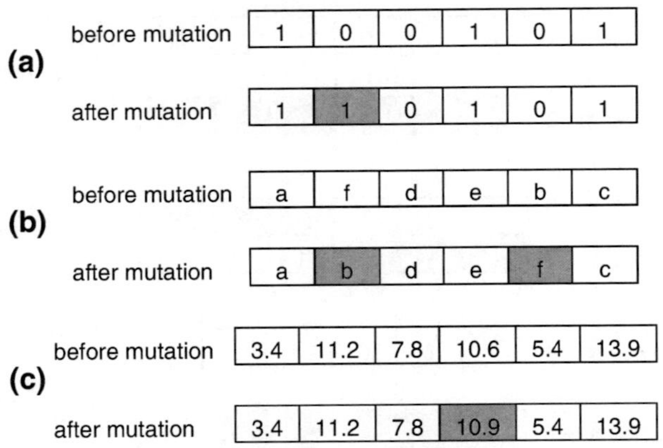

Fig. 4.11 Examples of mutation methods applied to different encoding. **a** binary encoding: the mutation operator inverts a randomly chosen bit (in this case the second bit is flipped from 0 to 1). **b** permutation encoding: two elements of the sequence are randomly selected and exchanged. **c** real value encoding: a gene is randomly selected and a small numerical quantity (mutation step) is added/subtracted

more objective function(s) and they are used as a measure of an individual's performance.

Whenever a set of objective functions, possibly in conflict with each other and/or measured in different units, has to be optimised (multi-objective optimisation problem—MOP), the optimum is generally defined as the solution giving all acceptable values for the objective functions and providing a good trade-off among them (Osyczka 2002). As a matter of fact the MOP solution search is articulated in two steps:

1. optimisation of the objective functions;
2. definition of the decision maker's criteria for selecting the solution to the problem.

Several techniques are available for managing the MOP process, according to the articulation of "decision" and "search" steps (Van Veldhuizen and Lamont 2000). Basically, three techniques are applied for solving MOP problems: optimisation of the highest priority objective, optimisation of an aggregated weighted sum of the objectives, and Pareto optimum search (Coello Coello et al. 2007).

By keeping track of the fitness value of the worst and the best individual at each iteration, the GA can implement stopping criteria based on fitness function values. For instance, the iterative procedure can be stopped if the fitness value of the best solution remains unchanged for a defined number of subsequent iterations, or if it reaches a defined optimum bound. Alternatively, the GA can stop when a maximum number of iterations is reached.

The following data have to be provided as input of the optimisation algorithm:

- GA parameters, which basically consist in crossover probability, mutation probability and population size;

- search domain limits, which are used to check actual feasibility of generated solutions;
- starting population, which can be either randomly generated or deterministically defined within a domain of acceptable solutions, according to the user's knowledge of the search space.

Several heuristic approaches have been proposed to find an optimal solution to the sensor placement problem. In Oh et al. (2007) a GA implementation is used with a multi-objective optimisation involving number, type and placement of sensors. The design objectives are related to coverage capabilities, network size, a weighted function referring to sensor types and proximity of target to sensors. A straightforward application of GAs provides a solution to a 2D unconstrained positioning problem. On the other hand, Ray and Mahajan (2002) apply GAs to the problem of placing receiving sensors, using a TDoA-based technique, within a ultrasound-based 3D positioning system. The design variables are the geometric locations of receivers whereas the objective is to avoid singularities and maximize measurement precision. A fixed number of sensors is considered. Both 2D and 3D configurations are analyzed. Realistic applications for navigating Autonomous Guided Vehicles (AGVs) within industrial environments have been approached through combinatorial optimisation methods. In Sinriech and Shoval (2000) the problem of cost minimization is addressed with a nonlinear optimisation model, where each sensor is characterized through its x–y coordinates in the planar space of the working environment (Sinriech and Shoval 2000). The aim is to adequately cover a set of critical points. Localisation precision is not taken into account. On the other hand, in Laguna et al. (2009) a combination of local search and tabu search is used to deal with a multi-objective optimisation problem. The proposed algorithm aims at maximizing coverage of a navigation area and measurement precision, minimizing the number of sensors employed. By assuming fixed vertical sensor positioning on the ceiling and fixed height of the receiving unit, the optimisation problem is actually restricted to a two-dimensional search. No blocking obstacles are assumed to exist in the communication range between transmitting and receiving devices.

A comparison between the performance of two heuristics (SA and GA), applied to the problem of deploying heterogeneous sensing devices under time-varying task requirements, is presented in Ramadan et al. (2007). Sensors differ as to lifetime, mobility capabilities, number of allowed moves (for the mobile ones), and reliability. The design objective consists in maximizing the area coverage. Although both methods provide satisfactory results, GA fares better than SA with the coverage objective.

Olague and Mohr (2002) apply a multi-cellular GA (Koza 1992) to the photogrammetric network design, i.e., the process of placing cameras for 3D measurements. The optimisation problem aims at finding camera positions and orientations which minimise the 3D point measurement error. GAs demonstrate that they are able to cope with high dimensional combinatorial optimisation problems related to discontinuities due to the presence of interposing obstacles.

4.5 The MScMS-I Framework: A Practical Case Study

The sensor positioning problem was solved within the MScMS-I framework through a software-assisted procedure, aimed at designing a flexible, efficient and modular sensor network by "pre-processing" technological, environmental and functional network-related issues. The "pre-processing" phase was structured into the following steps:

1. problem statement: definition of sensing model, working volume geometry, physical constraints, environmental operating conditions, localisation techniques;
2. task definition: setting of the measurement aim (e.g., single point coordinate measurement, distance measurement, shape reconstruction) and the measurement stations;
3. positioning strategy implementation: designing of the network of sensor nodes, according to either a regular or an optimal placement technique.

4.5.1 Problem Statement

4.5.1.1 Sensing Model

As reported in Chap. 3, the MScMS-I exploits a network of Cricket US transceivers, operating as transmitters, to spatially locate a portable hand-held probe, equipped with two identical US devices, operating as receivers. The radiation pattern characterizing Cricket transceivers is discussed in Sect. 7.2.1 and reported in Fig. 7.5. According to this pattern, reciprocal alignment and distance between transmitting and receiving devices represent key parameters of the communication and sensing model, affecting connectivity and coverage capabilities (Franceschini et al. 2010b). Hereafter, no distinction will be made between communication and sensing capabilities of US devices. The terms "field-of-sensing" and "communication volume" will be thus used as equivalent terms.

The sensing hardware was experimentally tested to obtain a mathematical model of the communication volume (see Fig. 4.12).

The model, which was implemented in the pre-processing software, was defined as a function of the following features:

- the "communication range" h, intended as the distance between the transmitting device (T) and the receiving devices (R$'$, R$''$), assuming that they are positioned in two parallel planes, facing each other;
- the "misalignment angle" (λ), defined by the maximum angle (in relation to the direction in which the transmitter faces) under which the communication between transmitter and receiver is maintained when the receiver is moved in any direction over the plane where it lies without changing its initial orientation.

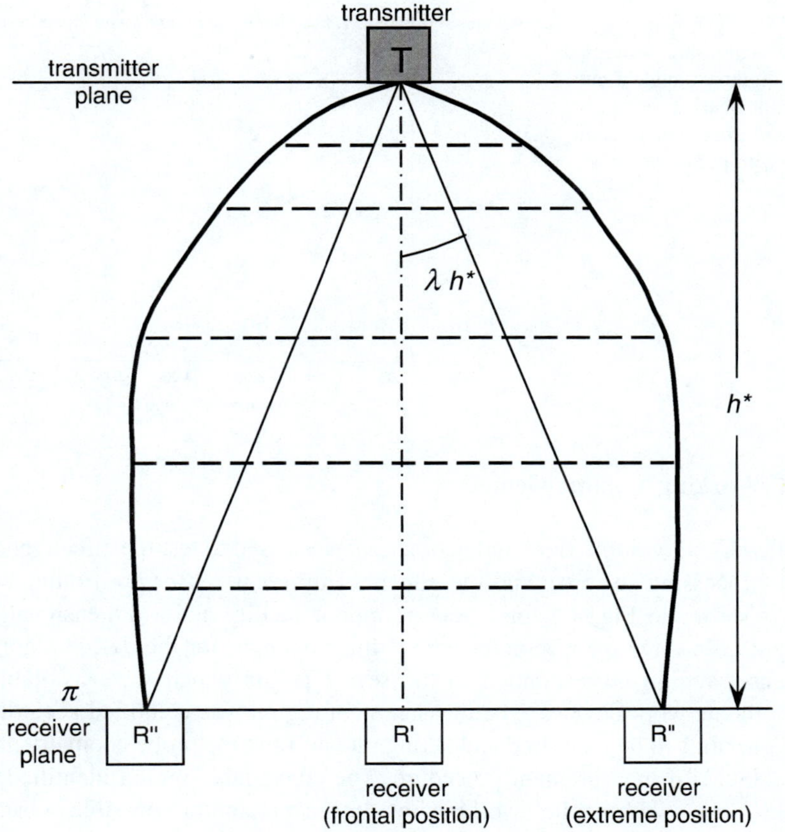

Fig. 4.12 Graphical interpretation of the "communication volume" (*bold lines*). For each communication range (h^*), intended as the distance between transmitter (T) and receiver (R′, R″) planes, the related misalignment angle (λ) allows the identification of the extreme receiver positions (R″) and, hence, a circular surface on the receiver plane. The actual communication volume, i.e., the region within which connectivity of US devices is guaranteed, is the envelope of the set of circular surfaces positioned between the transmitter plane and the reference plane π (Galetto and Pralio 2010). (with permission)

Experimental results showed a mutual dependence of these parameters (Franceschini et al. 2010). As a matter of fact, the misalignment angle monotonically decreases as the communication range increases, i.e., the receiver moves away from the transmitting node (see Fig. 4.13). This can be ascribed to the attenuation of the US signal (see Chap. 7 for further details). Therefore, the "field-of-sensing", intended as the real communication volume of US devices, has been defined as the envelope of a set of circular surfaces positioned between the transmitter plane and a reference plane (π in Fig. 4.12). Each surface may be identified as the intersection of a plane distant h^* from the transmitter and a cone which has the vertex in the transmitter centre and opening angle equal to $2\lambda(h^*)$.

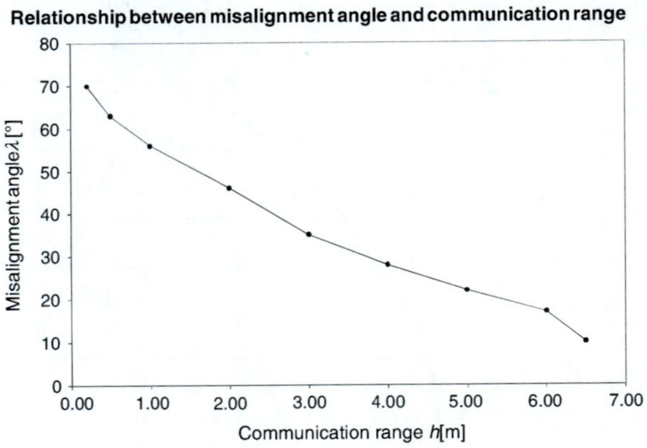

Fig. 4.13 Experimentally derived relationship between the misalignment angle λ and the communication range h (adapted from Galetto et al. 2011). (with permission)

4.5.1.2 Working Volume Geometry

Several working volume sizes and geometries were used to test the effectiveness of the pre-processing software and the effects of different sensor positioning strategies. As shown in Fig. 4.4 for a generic indoor facility and a dimensional measurement task, three main regions were defined as input for the design algorithm. The sensor region corresponded to the set of points which were available for placing the network devices. The measurement region was identified according to the set of points to be measured and taking into account the probe geometry as well as the MScMS-I measurement procedure. The unavailable region identified inaccessible areas within the working environment and/or possible obstacles. The information about position, size and shape of physical obstacles was used as input for the network design algorithm, in order to take into account shadowing effects.

4.5.1.3 Physical Constraints

Within the MScMS-I framework, there are no constraints related to the working volume geometry, as due to its network structure, MScMS-I works also correctly in environments with non-convex planimetry. Sensor devices, which are usually placed on the ceiling, can be positioned, if necessary, on the walls and the floor of the building. In specific cases, in order to "floodlight" shady areas, special trestles can be used to position sensors inside the working volume. Minimum distance between transmitters could be introduced as a design constraint in order to avoid mutual obstructions.

A constraint related to probe geometry was considered within this implementation. In fact, the distance between the two devices mounted on the measuring probe and the tip length are two basic elements for determining the extension of the connectivity areas for a given measured object (see Fig. 4.14). The network

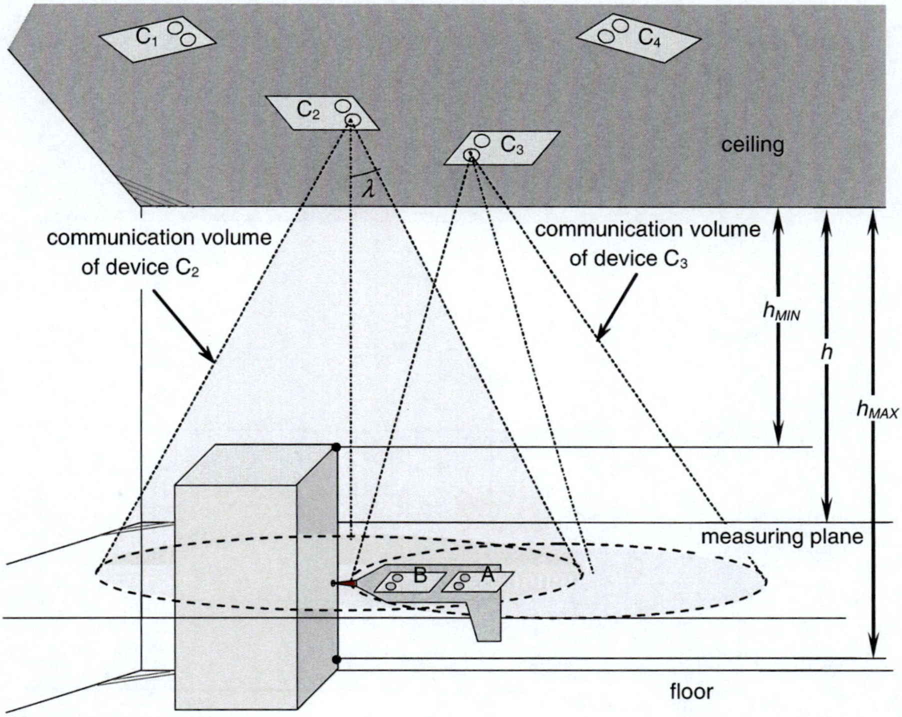

Fig. 4.14 Representation scheme of overlapping "communication volumes" of two US devices (in this case, C_2 and C_3) positioned on the ceiling. According to this simplified scheme, the actual communication volume of sensing devices (reported in Fig. 4.12) is represented as a cone. Each cone, identified by dashed lines, is a function of the relative distance h between the transmitting device C_i ($i = 1,\ldots4$) and a measuring plane, including the receiver devices A and B equipping the portable probe. Two reference distances, h_{MIN} and h_{MAX}, are also indicated, corresponding to the minimum and maximum distances, respectively, at which measurement is carried out (adapted from Galetto and Pralio 2010). (with permission)

has to ensure full coverage of the areas where the probe devices are positioned during measuring operations. This condition is not required for the tip. This results in an evident advantage when measuring complex surfaces, characterized by hollows or shady areas. In these cases special tips with particular geometries can be used in order to touch hidden points (Bosch 1995).

4.5.1.4 Localisation Techniques

A multilateration algorithm, based on measured TDoA of US signals, was implemented to spatially localise each probe device. According to this localisation technique, a minimum number of four transmitting devices was needed to guarantee point coverage capabilities (see Chap. 1).

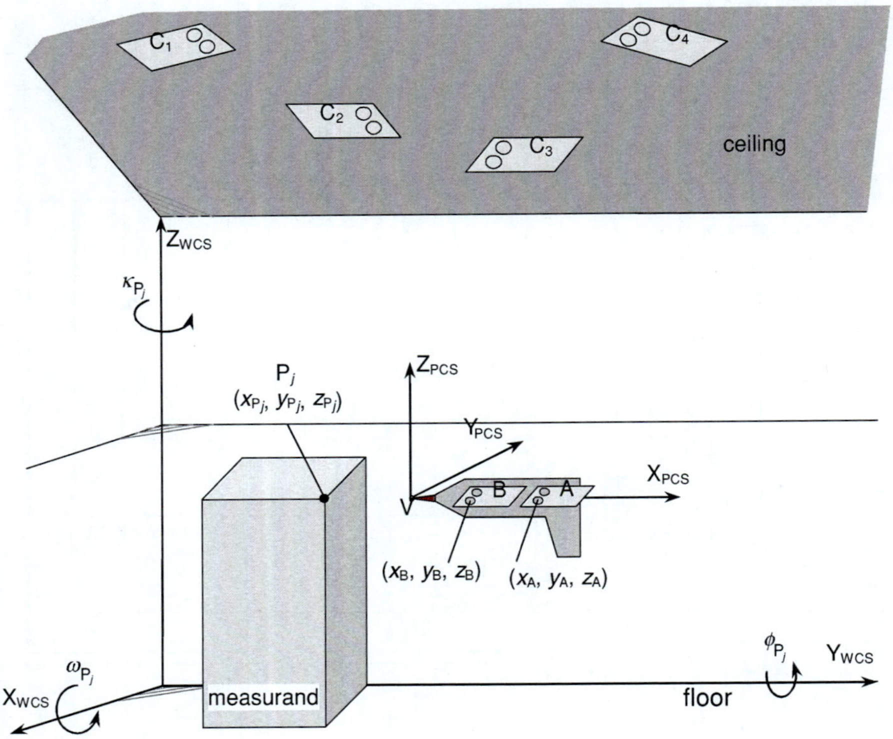

Fig. 4.15 Virtual reconstruction of the working setup. The measurement task is defined according to two coordinate reference frames. The world coordinate reference frame F_{WCS} has axes X_{WCS}, Y_{WCS}, and Z_{WCS}. The local (probe-fixed) coordinate reference frame F_{PCS} has the origin in the probe tip V, axes X_{PCS} and Y_{PCS} to form a plane containing the receiving devices A and B, and axis Z_{PCS} defined accordingly. The *black circle* represents a generic measurement point P_j located at coordinates (x_{Pj}, y_{Pj}, z_{Pj}) in the world coordinate reference frame. The probe orientation is defined through the set of rotations $(\omega_{Pj}, \phi_{Pj}, \kappa_{Pj})$ that, sequentially applied to F_{WCS}, align it to F_{PCS}. A measurement task where the probe is oriented upwards ($\phi_{Pj} = 0°$), perpendicular to the ceiling ($\omega_{Pj} = 0°$), and rotated of $\kappa_{Pj} = 90°$ with respect to the world coordinate reference system is represented

4.5.2 Task Definition

According to the MScMS-I application field, the measurement task aimed to characterize some geometrical features of differently shaped measurands, by providing distance measurements with a probe which had a fixed geometry. The measurement region was defined referring to the set of measurement points located on the surface of the reference object. A schematic representation of the working layout is provided in Fig. 4.15. The generic *j*th measurement point P_j was characterized by its spatial coordinates (x_{Pj}, y_{Pj}, z_{Pj}) in the world coordinate reference frame F_{WCS} and a set of angles $(\omega_{Pj}, \phi_{Pj}, \kappa_{Pj})$ that, sequentially applied to F_{WCS}, align it to the local (probe-fixed) coordinate reference frame F_{PCS}. These angles are thus related to the probe orientation with respect to F_{WCS}.

By defining the measurement task (i.e., the position of the probe tip and the probe orientation) and knowing the probe geometry, the position of receiving devices A and B in the 3D space can be calculated. It should be noted that, according to the system working principles, the region where sensor coverage has to be guaranteed is defined by the positions of the receiving devices on the probe, rather than by the measurement points located on the object. Although the coordinates (x_{Aj}, y_{Aj}, z_{Aj}) and (x_{Bj}, y_{Bj}, z_{Bj}) are used to generically define the position of devices A and B in the world coordinate frame F_{WCS}, it is worthy to observe that they refer to the centre of the receiving US transceivers equipping the probe Crickets (see Fig. 4.15).

4.5.3 Positioning Strategy Implementation

Two different sensor positioning strategies were investigated and experimentally tested within the MScMS framework. Both strategies carried out the network design according to the experimentally derived communication model of the US devices. Degree of coverage, measurement precision, and costs were addressed as objectives of the optimisation task and they were used as measures for the comparison between the proposed strategies.

The generic network device C_i was characterized through its spatial coordinates (x_{Ci}, y_{Ci}, z_{Ci}) in the world coordinate reference frame F_{WCS}, referring to the centre of the transmitting US transceiver (see Fig. 4.16).

The angular orientation of its field-of-sensing (communication volume) with respect to F_{WCS} was given through a set of rotations $(\omega_{Ci}, \phi_{Ci}, \kappa_{Ci})$ that, sequentially applied to the world coordinate reference system, align it to a local (C_i-fixed) coordinate reference system F_{CCS}.

A regular grid-based placement algorithm was implemented drawing a square configuration, characterized by a reference distance between sensor nodes. By constraining the devices to be placed on the ceiling, this strategy reduced the 3D positioning to a 2D problem, working on a fixed-altitude plane instead of the whole volume.

In order to improve the metrological performance of the distributed system, an optimal placement technique based on genetic algorithms was implemented. Although it cannot guarantee the global optimal solution, the heuristics-based approach was preferred to conventional search methods (enumerative as well as deterministic) due to the complexity of the problem. By dealing with a fully 3-dimensional working environment and measurement task definition, this approach was able to provide a more realistic and reliable network configuration.

4.5.3.1 Regular Grid-Based Strategy

Due to the sensor positioning constraints, a practical solution for solving the connectivity problem consisted in mounting the network devices on the ceiling of

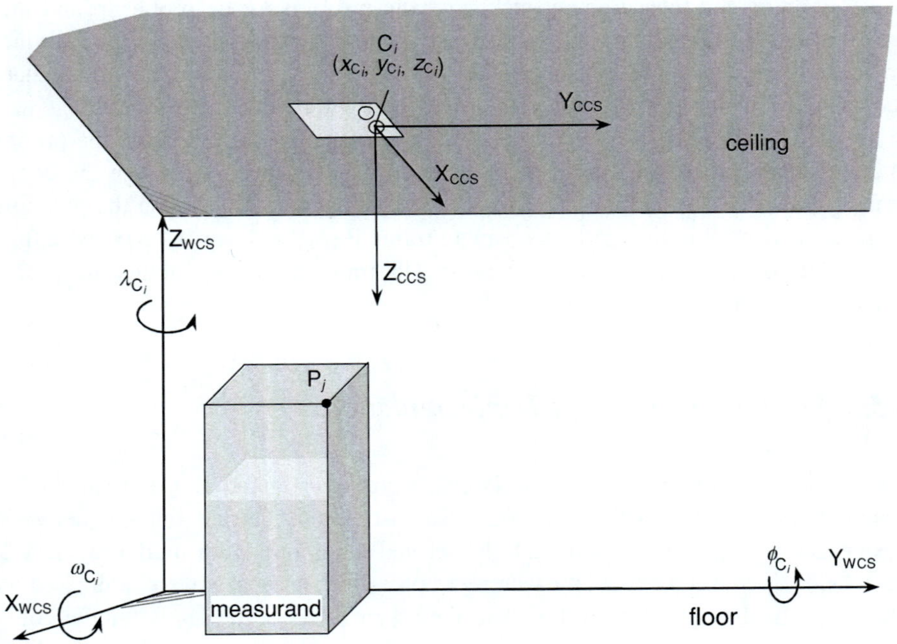

Fig. 4.16 Basic elements of the network device characterization. The generic device C_i is located at coordinates (x_{Ci}, y_{Ci}, z_{Ci}) in the world coordinate reference frame F_{WCS} with axis lines X_{WCS}, Y_{WCS}, and Z_{WCS}. The local (C_i-fixed) coordinate reference frame F_{CCS} has the origin the centre of the transmitting US transceiver, axes X_{CCS} and Y_{CCS} parallel to the US device surface, and axis Z_{CCS} perpendicular to the plane $X_{CCS}Y_{CCS}$ and positive in the sensing direction. The black circle represents a generic measurement point P_j located at coordinates (x_{Pj}, y_{Pj}, z_{Pj}) in the world coordinate reference frame. The device orientation is defined through the set of rotations (ω_{Ci}, ϕ_{Ci}, κ_{Ci}) that, sequentially applied to F_{WCS}, align it to F_{CCS}. A layout where the device C_i is oriented downwards ($\phi_{Ci} = 180°$) and parallel to the ceiling ($\omega_{Ci} = 0$) is represented

the working environment, as shown in Fig. 4.14. Measurements were assumed to be performed by orienting the probe Crickets upwards. In this condition, the distance between the ceiling and the reference plane, at the maximum distance from the ceiling at which it was planned to work, was assumed to be "maximum communication range" (h_{MAX}). The corresponding "misalignment angle" was called "minimum misalignment angle" (λ_{MIN}). Furthermore, it was assumed to "floodlight" only the zones where probe Crickets are positioned during the measurement (according to the object shape, the probe geometry and the measurement procedure). Under these conditions, the concept of beacon "density" was defined as the number of network devices that should be placed per unit of surface on the ceiling, in order to correctly "floodlight" a given region of an horizontal plane positioned at a generic distance h from the ceiling (see Fig. 4.14). According to the schematic representation of Fig. 4.14, the covered area is determined by the circular surfaces representing the intersection of the horizontal measuring plane and the cones generated by the network devices (see Fig. 4.17).

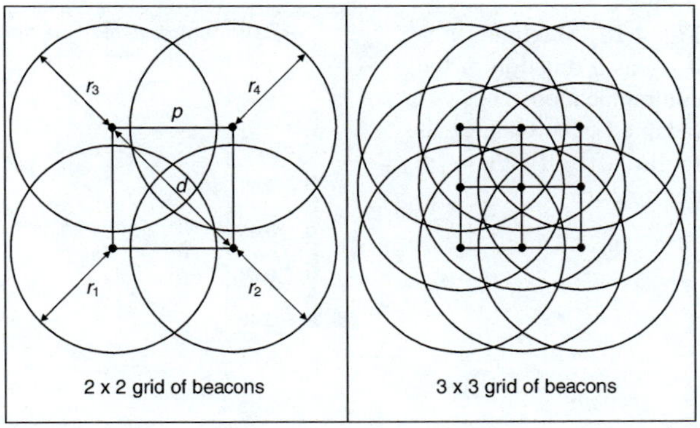

Fig. 4.17 Covered areas versus beacon "density" at a given distance h from the ceiling (Franceschini et al. 2008). (with permission)

A square mesh grid arrangement was selected. Referring to Fig. 4.17, it can be observed that a circle, having radius r_h on the plane at a distance h from the ceiling and centred on a corner of the square mesh having diagonal d, overlaps the opposite corner if $d \leq r_h$. Defining as pitch (p) the distance between two nodes of the grid, the diagonal of the mesh is given by:

$$d = \sqrt{2} \cdot p \tag{4.2}$$

Assuming that all the circles generated on a plane have the same extension ($r_h = r_1 = r_2 = r_3 = r_4 = \ldots$) and using the experimentally derived relationship in Fig. 4.13 to relate the misalignment angle λ and the distance h, the pitch can be rewritten as:

$$p \leq \frac{1}{\sqrt{2}} \cdot h \cdot \tan(\lambda(h)) \tag{4.3}$$

Considering that a minimum number of four devices/distances are needed to localise a device using a multilateration technique, the maximum pitch can be defined as:

$$p_{MAX} = \frac{1}{\sqrt{2}} \cdot h \cdot \tan(\lambda(h)) \tag{4.4}$$

According to the λ–h relationship shown in Fig. 4.13, its variation with the communication range h is a non-monotonic function (see Fig. 4.18).

Provided that the measuring volume ranged from a minimum distance h_{MIN} to a maximum distance h_{MAX}, a conservative pitch value p_{GRID} was estimated as:

$$p_{GRID} = \min\left(\frac{1}{\sqrt{2}} \cdot h_{MIN} \cdot \tan(\lambda(h_{MIN})), \frac{1}{\sqrt{2}} \cdot h_{MAX} \cdot \tan(\lambda(h_{MAX}))\right) \tag{4.5}$$

Due to the behaviour of λ–h relationship, this pitch size guaranteed the complete coverage of the measuring planes at h_{MIN} and h_{MAX}, as well as of the enclosed volume.

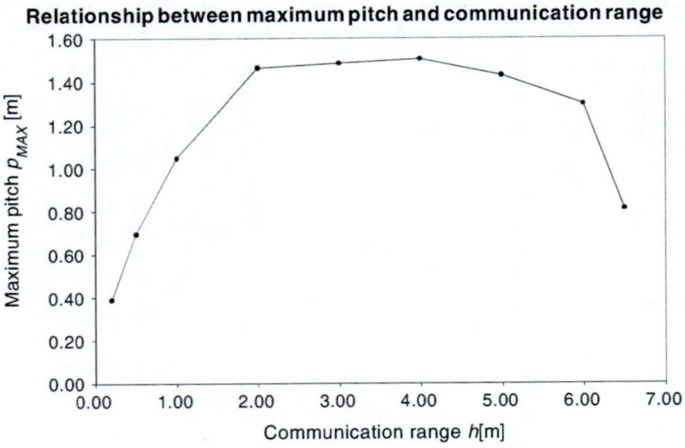

Fig. 4.18 Maximum pitch p_{MAX} as a function of the communication range h (adapted from Galetto and Pralio 2010). (with permission)

4.5.3.2 GA-Based Strategy

Evolutionary algorithms, and specifically genetic algorithms, were selected according to the results of an explorative analysis of widely used search techniques (such as simulated annealing, tabu search, scatter search). The evolutionary technique, applied to the present multi-objective optimisation problem, was able to provide a comprehensive representation of the 3D search space and to limit the computational burden of the searching phase. Each individual, as a possible solution of the optimisation problem, was represented by a configuration of distributed sensor nodes $\xi = (C_1, C_2, \ldots, C_{n_{max}})$. The external parameters (x_{Ci}, y_{Ci}, z_{Ci}, ω_{Ci}, ϕ_{Ci}) were selected as decision variables of the optimisation problem (see Fig. 4.16). Since, due to its axial symmetry property, variations of the κ_{Ci} orientation angle do not cause changes to the communication volume, a number of decision variables $n_{dv} = 5$ was defined. A binary encoding representation was used.

Selection of parents for reproduction was made according to a tournament selection method, by randomly choosing two individuals and selecting the candidate parent with the best fitness value. A steady state strategy was implemented, i.e., at each iteration the new population was formed referring to the parents and the newly generated offspring. One-point crossover and bit mutation were applied.

The parameter n_{max}, i.e., the maximum number of available sensors, was introduced by the user through the initial configuration and represented an upper limit for the network sizing. The population thus consisted of a $n_p \times n_{max} \times n_{dv}$ multi-dimensional array, where n_p is the number of individuals per population. The sensor region, i.e., the set of candidate positions for sensor placement, was bounded according to the working environment dimensions and physical constraints. Network configurations characterized by different sensor numbers ($\leq n_{max}$) were evaluated by enabling/disabling nodes.

A criterion based on the maximum number of iterations was implemented to stop the iterative procedure.

Three objective functions were considered in the optimisation procedure, taking into account, at each iteration, overall cost, coverage capabilities, and measurement precision.

As reducing the number of sensor nodes represents an important issue in order to build a flexible system, the cost function (O_1) was defined in terms of number of enabled sensors $(n_{act} \leq n_{max})$ as follows:

$$O_1 = 1 - \frac{n_{act}}{n_{max}}; \quad O_1 \in [0, 1] \tag{4.6}$$

This function represents the cost reduction obtained by the current layout (using n_{act} sensors) compared to the initial reference configuration (using n_{max} sensors). The lower limit of the definition range corresponds to a configuration using all the available sensors $(n_{act} = n_{max})$ whereas the upper limit characterizes an empty sensor set.

It is noteworthy that taking into account cost reduction as a key objective of the optimisation process has a twofold motivation. Firstly, reducing the network size entails a reduction in direct costs (price per sensor unit), labour costs, and setup times. Since sensor positioning and calibration are time-consuming activities, labour costs per placed unit represent an additional factor that should be taken into account whenever economic impact is evaluated. Furthermore, as the proposed optimisation strategy could be applied to sensing devices other than ultrasound-based (e.g., optical sensors), by simply changing the communication model, the cost of a single sensor unit could become a serious constraint.

The coverage capabilities of different sensor configurations were stated, for each measurement point P_j, as the number of sensors n_{covj} including the receiving devices within their communication volumes, under the assumption that transmitting and receiving devices were positioned in parallel planes. The effects of object shadowing on communication cones were taken into account in the coverage test. Since US transducer performance is affected by obstacles interposing between receiving and transmitting devices, the actual geometry of the measured object was modified according to its shadow. As a matter of fact, the shadow of the measured object was subtracted from the "ideal" communication volume, reducing the covered area. The coverage function (O_2), intended as a "measure" of network coverage capabilities, was related to the number of transmitting devices having each measurement point P_j $(j = 1,\ldots,m)$ within their sensing range. Therefore, it was algorithmically defined as:

$$O_2 = \frac{\sum_{j=1}^{m} O_{2_j}}{m}; O_2 \in [0, 1] \tag{4.7}$$

where

$$O_{2j} = \begin{cases} 1 & \text{if } \min\left(n_{\mathrm{cov_{j,A}}}, n_{\mathrm{cov_{j,B}}}\right) \geq n_{min} \\ \dfrac{\min\left(n_{\mathrm{cov_{j,A}}}, n_{\mathrm{cov_{j,B}}}\right)}{n_{min}} & \text{otherwise} \end{cases} \tag{4.8}$$

and n_{min} is the minimum number of transmitting sensors needed to locate the measuring probe (degree of coverage).

The lower limit of the definition range describes a layout in which no communication is established between transmitting and receiving devices for any measurement point. On the other hand, the upper limit characterizes an operating condition in which all the measurement points can be adequately localised as they are all covered by the minimum number of sensors.

The network efficiency was evaluated through the measurement precision. A Dilution of Precision (DOP)–based method was used to give a measure of the quality of the network geometry. The concept of dilution of precision is applied to GPS as well as to other distributed systems using distance measurements from reference points to solve a localization problem (Bar-Shalom et al. 2001; Hofmann-Wellenhof et al. 2001). It specifies the additional multiplicative effect of the network geometry on the measurement precision. For determining the precision in positioning the receiving devices, the precision in distance measurement combines with the relative position of the network nodes. A strong geometry, corresponding to an efficient network, is characterized by low DOP values. On the contrary, when the nodes are close to each other the geometry is weak and the DOP value is high. Various expressions for the DOP are given depending on the specific components of interest, such as the three-dimensional receiver position coordinates (PDOP—Position DOP), the horizontal coordinates (HDOP—Horizontal DOP), or the vertical coordinate (VDOP—Vertical DOP). According to the present application, the PDOP was used as a measure of the network efficiency. As reported in Eq. 4.9, the PDOP associated with the measurement of a point P_k, adequately covered by the sensor network ($O_{2k} = 1, k \leq m$), was mathematically derived from the spatial coordinates (x_{Ci}, y_{Ci}, z_{Ci}) of the generic network node C_i used in the localisation problem, the spatial coordinates ($x_{Pk,j}$, $y_{Pk,j}$, $z_{Pk,j}$) of the jth receiving device (j = A,B) corresponding to kth measurement point P_k, and their relative distances $d_{i,k,j}$:

$$PDOP_{k,j} = f\left(x_{C_1}, y_{C_1}, z_{C_1}, \ldots, x_{C_N}, y_{C_N}, z_{C_N}; x_{P_{k,j}}, y_{P_{k,j}}, z_{P_{k,j}}; d_{1,k,j}, \ldots, d_{N,k,j}\right) \quad (4.9)$$

We refer the reader to Sect. 8.6 for a detailed discussion about the mathematical formulation of the PDOP term.

The precision function (O_3), intended as a "measure" of the quality of network geometry, was thus defined by comparing the $PDOP_{k,j}$ to a user-defined upper bound value ($PDOP_{lim}$) as follows:

$$O_3 = \frac{\sum_{k=1}^{m_{cov}} O_{3_k}}{m_{cov}}; O_3 \in [0, 1] \quad (4.10)$$

where

$$O_{3_k} = \begin{cases} 1 - \dfrac{\max\left(PDOP_{k,A}, PDOP_{k,B}\right)}{PDOP_{lim}} & \text{if } \max\left(PDOP_{k,A}, PDOP_{k,B}\right) \leq PDOP_{lim} \\ 0 & \text{otherwise} \end{cases} \quad (4.11)$$

and m_{cov} is the number of measurement points characterized by unit coverage function O_2. The lower/upper limit of the definition range refers to a network layout characterized by unacceptable/acceptable PDOP values for all the measurement points.

The fitness function (FF), expressing the need for minimizing the network cost and maximizing sensor coverage as well as measurement precision, was defined as a linearly aggregated weighted sum of the objectives, according to a Conventional Weight Aggregation (CWA) approach (Jin et al. 2001):

$$FF = K_1 O_1 + K_2 O_2 + K_3 O_3 \tag{4.12}$$

where the weighting coefficients ($0 \leq K_i \leq 1$; $\Sigma K_i = 1$) represent the relative importance given by the user to the problem objectives. It should be noted that the results of the multi-objective optimisation problem are strongly related to the definition of the fitness function, which entails modelling the relationships among the objectives, and the selection of weights, which implies to quantify the relative importance of the objectives.

Several tests were carried out to evaluate the performance of the two positioning strategies (Galetto and Pralio 2010). Differently sized and shaped measurands as well as a variety of measurement procedures (i.e., positions and orientations of the receiving devices) were implemented in the pre-preprocessing software.

The GA-based strategy demonstrated to outperform the regular grid-based technique as to cost reduction capabilities, besides competitive performance as to sensor coverage and measurement precision. Moreover, the regular positioning approach, being unable to take into account the presence of a physical object between the measuring planes (at h_{MIN} and h_{MAX}), showed an expected performance degradation when shadowing effects were considered.

Interesting considerations regarding potentialities of the GA-based strategy resulted from evaluating its performance with fully 5-DoF sensor characterization. The GA-based approach effectively demonstrated capabilities to further improve its performance according to spatial and rotational constraints of sensor positioning.

Chapter 5
System Calibration

5.1 Concepts

As specified in the International Vocabulary of Metrology, the calibration is the "operation establishing the relation between quantity values provided by measurement standards and the corresponding indications of a measuring system, carried out under specified conditions and including evaluation of measurement uncertainty" (JCGM 100:2008 2008).

Historically, the term "calibration" probably was first associated with the precise division of linear distance and angles using a dividing engine[1] and the measurement of gravitational mass using a weighing scale.[2] These two forms of measurement alone and their direct derivatives supported nearly all commercial and technological development from the earliest civilizations until the Industrial Revolution. At that time, the presence of different standards of reference (the units of measurement could have been different from kingdom to kingdom or from region to region) had already highlighted the importance of calibration of measuring instruments.

Later on, with the Industrial Revolution the concept of calibration assumed an even greater importance. The Industrial Revolution, in fact, introduced wide scale use of indirect measurement. Indirect measuring instruments are instruments that do not directly measure the quantity of interest, but basic quantities from which one of interest is derived. As an example, consider the spring balance presented in Fig. 5.1. This tool does not directly measure the mass of the object placed on its pan. It measures rather the travel of the piston on which the object to be measured is placed over. The travel of the piston is proportional to the force exerted by the object and is therefore proportional to its mass. But how to determine the relation between the piston excursion and the mass of the object to be measured? This is

---

[1] A dividing engine is a device specifically employed to mark graduations on measuring instruments (Daumas 1989).

[2] A weighing scale, in general, is a measuring instrument for determining the weight or mass of an object (Daumas 1989).

F. Franceschini et al., *Distributed Large-Scale Dimensional Metrology*,
DOI: 10.1007/978-0-85729-543-9_5,

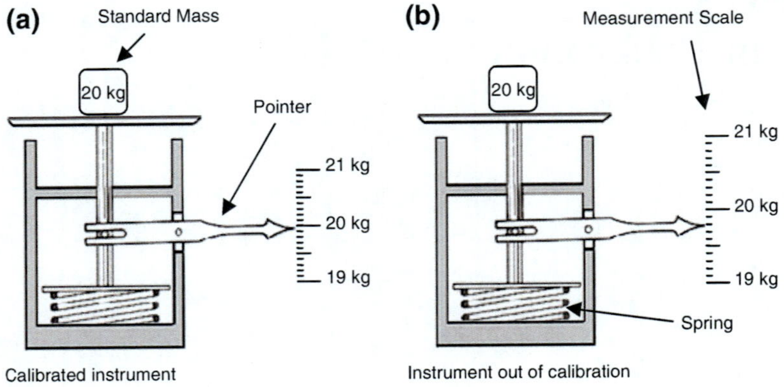

Fig. 5.1 An example of a spring balance. In the figure on the *right* (**b**), due to a poor calibration, the scale associated with the calibration is incorrect. This causes a misreading of the measurement results of the instrument. In the figure on the *left* (**a**), the scale associated with the calibration and thus the indication provided by the instrument are correct

precisely the purpose of a calibration that, through the use of some reference standards, defines a correspondence between the indication of the instrument and the nominal value of the measured quantity. In the example, this means defining an appropriate measurement scale.

Figure 5.1 compares the behaviour of the same instrument operating under different conditions. If, for some reason (such as drift, environment conditions, electrical supply, wear of components, process changes, etc.), the instrument runs out of calibration then it provide erroneous results.

For this reason, and because of its importance, the calibration is typically performed before the use of any metrology system.

For distributed systems, which are composed by several distributed metrological devices, this stage becomes crucial. The measurement indications provided by each metrological device need to be aligned on the same reference. It is clear how errors made at this stage may strongly affect the subsequent phase of measurement.

This chapter is organized as follows. Sections 5.2 and 5.3 are dedicated to a general description of the calibration procedure. Common traits of the calibration procedures of different distributed metrology systems are highlighted and discussed. As an example the calibration procedures implemented for MScMS-I are gone in more depth in Sect. 5.5. Finally Sects. 5.6 and 5.7 provide some hints about the calibration procedure for MScMS-II and iGPS™ respectively.

5.2 The Goal of Calibration

Based on different technologies, metrology instruments have different calibration procedures aimed at estimating different unknown parameters. Generally, the parameters considered during the calibration phase can be classified into two categories:

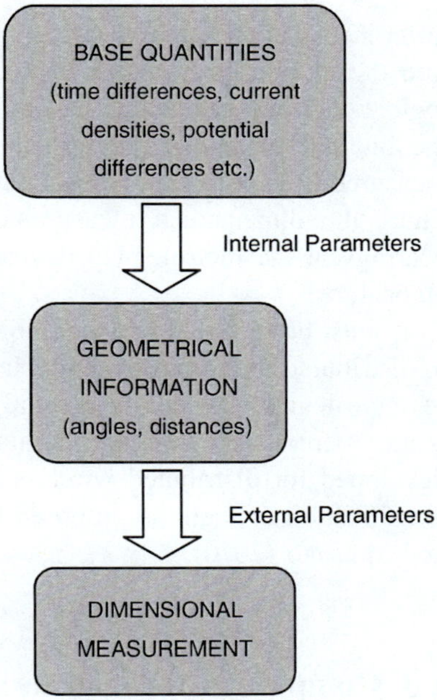

Fig. 5.2 From base quantities to dimensional measurement. By knowing both the internal and external parameters it is possible to translate the measured base quantities into geometrical information

- *Internal (intrinsic) parameters* These parameters are directly related to each metrological device. Internal parameters are specifically related to the technology. Examples of internal parameters can be the focal length or the skew/distortion coefficients of CCD sensors.
- *External (extrinsic) parameters* These parameters are not related to the technology of the metrological devices, they rather depend on external/environmental factors that can affect the indications of the metrological devices. Examples of external parameters are the temperature, pressure, humidity of the working environment or the position and orientation of the metrological devices.

The estimation of the values of these parameters is crucial because the dimensional measurement is derived by each distributed metrology system through a nonlinear function of these parameters. In fact, the metrological devices of a distributed system do not directly provide geometrical or dimensional measurements, they rather evaluate base quantities[3] such as differences in timing between pulses of different origins, current densities, or potential differences. The knowledge of the internal parameters related to the each metrological device allows to translate the measured base quantities into geometrical information. Only by knowing the external parameters, the geometrical information can be translated into dimensional measurements (see Fig. 5.2).

In order to better clarify the concept of internal/external parameters let us consider the example of the iGPS™. In this case the base quantities are the time

[3] The term "base quantity" is defined by JCGM as the "*quantity, chosen by convention, used to define other quantities*" (JCGM 100:2008 2008).

differences between the electrical pulses produced by the photodiodes embedded into the mobile probe when lighted by the laser beams emitted by each metrological device. By knowing the geometry of each metrological device (specified by the internal parameters described in Sect. 5.7) the system is able to turn the measured time pulses into an azimuth and elevation angle (see Chap. 2, Sect. 2.2). Then, the dimensional measurements are derived knowing the positions and bearings of the metrological devices (external parameters) through triangulation algorithms.

It must be said that the operation of external parameters calibration is peculiar to distributed metrology systems. In particular, among all parameters, the position of metrological devices is certainly one of the most important. Regarding this point, there is a extensive literature about localisation algorithms specifically developed for distributed wireless sensor networks. Section 5.4 contains a brief classification of them as proposed by Franceschini et al. (2009), Mastrogiacomo and Maisano (2010).

5.3 Common Approach to System Calibration

Calibration procedures implemented by distributed systems are similar to some extent. They are typically organized into two phases:

- *Calibration Data Collection* During this phase, some reference artefacts (or standards) are introduced. At this stage the measurement system—although not calibrated—is typically used to gather a collection of measurements of the geometrical features of the reference artefacts.
- *Calibration Data Computing* The a priori knowledge of the geometrical features of the reference artefacts allows the estimation of the unknown internal and external parameters. This takes place during the second phase, which is usually performed off-line. In this phase the system searches for those parameters values that minimize the deviations of the measurement data gathered during the first phase, from their a priori known reference values.

5.3.1 Data Collection

As stated above, during this phase some reference artefacts have to be introduced. The reference artefacts are objects with known dimension made of special materials (such as composite materials) that are almost insensitive to variation of environmental parameters (temperature, pressure, humidity, light conditions and so on). Typically their geometrical features are univocally evaluated using high accuracy metrology instruments such as CMMs or interferometers and thus they can be used as reference to calibrate the metrological devices.

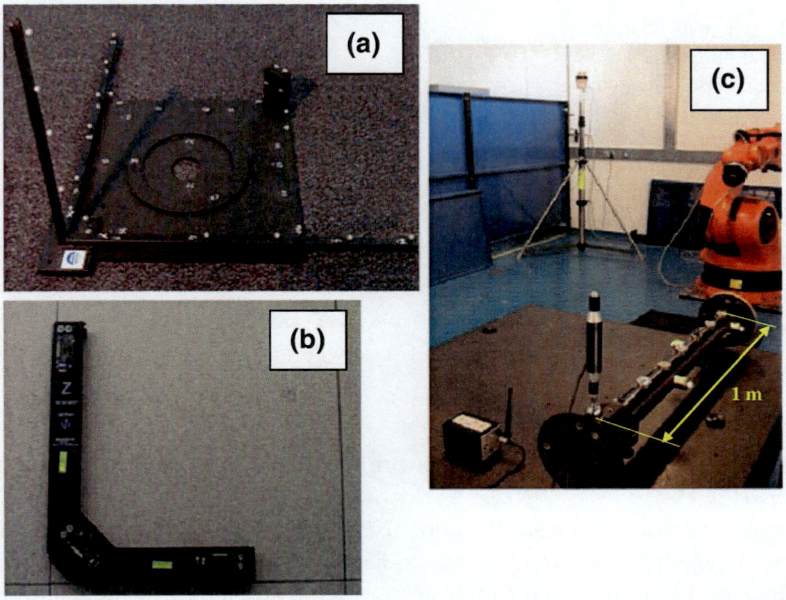

Fig. 5.3 Examples of reference artefacts. **a** Typical reference used for the calibration of optical distributed dimensional metrology systems. **b** One of the reference artefacts used for the calibration of MScMS. **c** The vector bar used for the calibration of the iGPS^{TM} (with permission of Bath University Mechanical Engineering Laboratories)

Figure 5.3 presents some dimensional reference artefacts. In detail, Fig. 5.3a shows a typical reference artifact used for the calibration of optical distributed metrology systems. Figure 5.3b shows one of the reference artefacts used for calibration of MScMS. In Fig. 5.3c is shown the vector bar used for the calibration of the iGPS^{TM}.

During the collection phase these objects are moved within the working volume in different positions and orientations and repeatedly measured. Since, at this stage, the distributed system is not yet calibrated, the measurements collected and stored during this phase are not actually dimensional measurements, but just the base quantities provided by each metrological device. This information is then processed during the data computing phase.

Depending on the distributed system, the reference artefacts may have also the role of defining a global coordinate reference system.

5.3.2 Data Computing

This is the core part of the calibration procedure. The goal of this phase is to find that estimation of the parameters which better aligns the measurements recorded

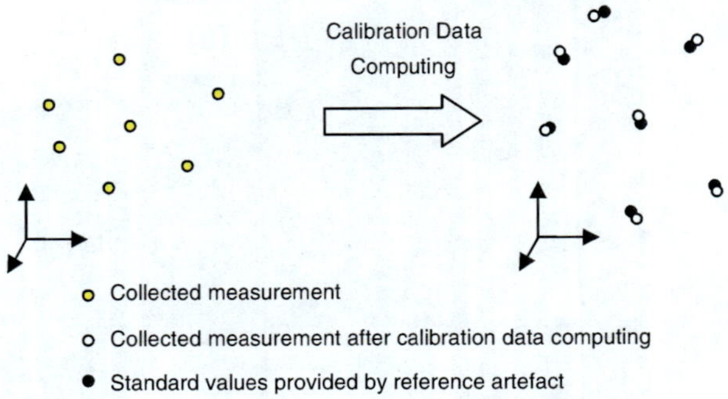

Fig. 5.4 Schematization of the Calibration Data Computing phase. The goal of this phase is to find the better alignment between collected measurements and reference values provided by reference artefacts

during the collection phase with the a priori known dimension of the reference artefacts (Fig. 5.4).

To this purpose a Fitness Function $FF(\chi)$ is typically defined. The fitness function (sometimes referred to as "objective function" or "cost function") describes the fitting of a mathematical model for measurement procedure to the experimental measurement data. It is a function of the set of unknown internal and external parameters (χ) that are believed to significatively affect the measurement phase. Typically it has the following properties:

- $FF(\chi) \geq 0$ for any set of parameters;
- $FF(\chi)$ is nonlinear in χ;
- $FF(\chi)$ has a global minimum in χ_0, which is that particular set of parameters values able to align measurements provided by the instrument with "the quantity values provided by reference artefacts".

Thus the calibration problem can be seen as a nonlinear optimisation problem aimed at finding the values of parameters χ minimizing the fitness function:

$$\min_{\chi}(FF(\chi)) \tag{5.1}$$

The solution of this problem can be iteratively found from a first approximation of the parameters values. Most often, but not necessarily, this results in minimising the sum of the squares of the deviations of the measurement data from their values predicted with a non-linear function of the unknown parameters (Triggs et al. 2000; Lourakis and Argyros 2009). A range of general purpose optimisation algorithms, such as for instance Gauss-Netwon and Levenberg–Marquardt rather than simulated annealing or can be used to minimize the non-linear objective function. Alternatively, increased efficiency can be gained if these algorithms are adjusted to account for the sparsity of the matrices arising in the mathematical description of 3D reconstruction problems (Lourakis and Argyros 2009). In any case it is remarkable that, whatever it is, the choice of optimisation algorithm is strongly related to the problem definition and thus may vary according to it.

5.4 Localisation Algorithms

A great abundance of localisation algorithms are described in literature. Generally, they are designed to be applied to a typical sensor network, consisting of a large number of nodes with a dense distribution. As a consequence, many of them do not fit to small networks, with few distributed nodes. According to their features, localisation algorithms can be classified within many categories.

The first categorisation is based on the presence (or absence) of nodes with pre-configured coordinates:

- *Anchor-based algorithms* The localisation is implemented by selecting a set of reference nodes ("landmarks", "anchor-nodes") with known coordinates (see Fig. 5.5). A localisation system, with "anchor-nodes", has the limitation that it needs another localisation system (e.g., GPS) to determine the anchor-nodes positions. Furthermore, a large number of anchor-nodes are required, for the resulting position errors to be acceptable (Priyantha et al. 2003).
- *Anchor-free algorithms* They use local distance measurements among nodes to determine their respective coordinates. They do not assume the availability of nodes with pre-configured coordinates (see Fig. 5.6).

The second categorization is based on the way node localisations "propagate" in the network:

- *Incremental algorithms* These algorithms usually start with a set of three or more reference nodes with known coordinates. Other nodes in the network can contact the reference nodes and determine their own coordinates. As an unknown position node obtains a acceptable position estimate, it may serve as a new reference point. This process can be incrementally applied until all nodes in the network have obtained their coordinates.
- *Concurrent algorithms* In this approach, many pairs of sensors communicate and share measurements, in order to achieve localisation for all sensors. Rather than solving each sensor position one at time, all sensor positions are simultaneously estimated. Such localisation systems not only allow unknown-location devices to make measurements with known-location references, but they additionally allow unknown-location devices to make measurements with other unknown-location devices. The additional information gained from these measurements between pairs of unknown-location devices enhances the accuracy and robustness of the localisation system. Such systems have been also referred as "cooperative" (Patwari et al. 2005).

The third categorization is based on computational distribution:

- *Centralized algorithms* Computing is performed by a single centralized node or network device. All nodes broadcast information to a single computer to solve the localisation problem.

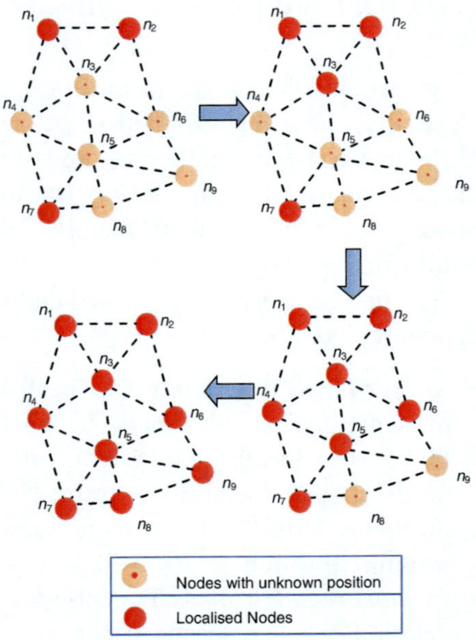

Fig. 5.5 An example of anchor based incremental algorithm. In the example n_1, n_2 and n_7 act as anchor nodes. All the other nodes localise themselves incrementally according to the positions of the already localised nodes

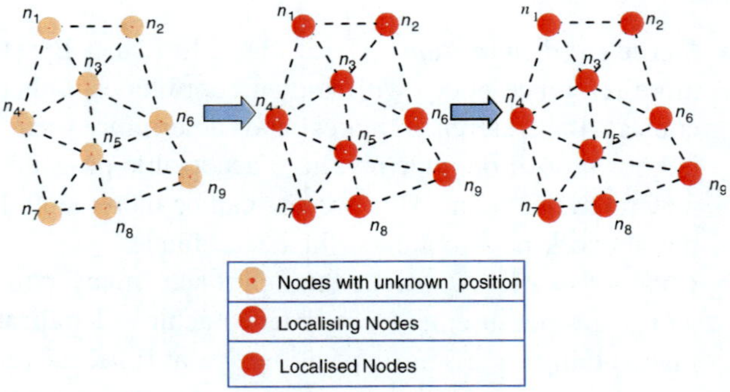

Fig. 5.6 An example of anchor free localisation algorithm. All nodes localise themselves concurrently basing on distance information

- *Distributed algorithms* Computing is equally distributed among network nodes. Each node receives location information from neighbouring nodes, performs computation, and retransmits the obtained results to them.

It must be said that the algorithms for the calibration of the external parameters of distributed metrology systems are inspired by the algorithms for the localization of wireless sensor networks. Nevertheless, these techniques cover only partially the problem of calibration of these systems which, in order to function correctly, commonly need the knowledge of other (internal and external) parameters in addition to positioning metrological devices.

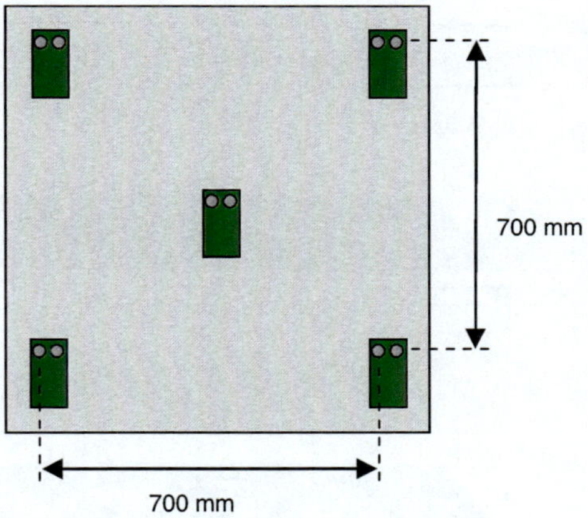

Fig. 5.7 Dimensional reference artefact used for the calibration of MScMS-I. This artifact embeds five devices placed at a mutual distance of 700 mm

5.5 Calibration Procedures for MScMS-I

This section presents some procedures implemented to face the calibration of MScMS-I. The three proposed procedures are aimed just at the calibration of a subset of the external parameters i.e., the positions of the metrological devices (Crickets).

All the procedures rely on the use of a dimensional reference artefact with known geometry, which embeds a sufficient number of network devices and can be easily moved within the measuring volume (Fig. 5.7).

The procedures are compared considering the network features they can deal with, the computational workload they require and their major advantages and weaknesses. The position of the network devices has been compared to a nominal position achieved using a laser tracker, whose nominal accuracy is greater than the MScMS-I accuracy (see Sect. 2.4.2).

For all the procedures, the fitness function is given together with the logic behind its definition.

5.5.1 First Procedure

The key idea of this procedure is to place the reference artefact in known positions (see Fig. 5.8). Each time the reference artefact is moved, it is kept still for a while. During the time period in which it is not moving, the reference artefact is able to communicate with the network devices and estimate their distances from the embedded devices. Knowing these distances and the position of the reference artefact in a global reference coordinate system, the network devices can be localized by solving the optimisation problem described in the following.

Let M be the number of wireless devices embedded in the reference artefact and N the number of network device to be localized. As said before, the reference

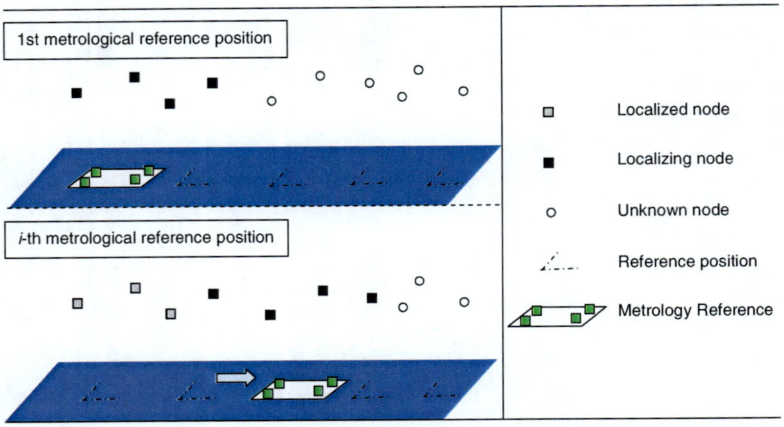

Fig. 5.8 Schematic representation of the first localisation procedure. The multiple positioning of the reference artefact progressively localizes the network nodes

artefact is placed in different positions under the network devices. Let P be the number of reference artefact positioning. Since for each repositioning, the position of the reference artefact is supposed to be known, the reference artefact defines $M \cdot P$ reference points (R_j) with known coordinates ($\left(x_{R_j}, y_{R_j}, z_{R_j}\right)$, with $j = 1, \ldots, M \cdot P$). Let also C_1, ..., C_N be the points with unknown coordinates $(x_{C_i}, y_{C_i}, z_{C_i})$, with $i = 1$, ..., N corresponding to the network devices to be localized.

During the phase of the reference artefact repositioning, the wireless devices are able to measure the distance $\left(\tilde{d}_{C_i,R_j}\right)$ between the points defined by the reference artefact (R_j) and the points corresponding to the network devices (C_i).

Knowing the set of measured distances, it is possible to define a distance vector ($d_{i,j}$ with $i = 1,\ldots,N$ and $j = 1, \ldots, M \cdot P$) associated to each network node (C_i) as:

$$d_{i,j} = \begin{cases} \tilde{d}_{C_i,R_j} & \text{If network device } C_i \text{ can estimate its distance to } R_j \\ 0 & \text{Otherwise} \end{cases} \tag{5.2}$$

As a consequence let's define the connection set I_i as the set of reference points to which the network device (B_i) is able to estimate the distance:

$$I_i = \left\{ j \in \{1, \ldots, M \cdot P\} : d_{i,j} \neq 0 \right\} \tag{5.3}$$

The unknown position of the ith network device ($B_i \equiv (x_{B_i}, y_{B_i}, z_{B_i})$) can be found as the position that, for each $j \in I_i$, minimizes the difference between Euclidean and measured distance:

$$\left(\sqrt{(x_{C_i} - x_{R_j})^2 + (y_{C_i} - y_{R_j})^2 + (z_{C_i} - z_{R_j})^2} - \tilde{d}_{C_i,R_j} \right)_{j \in I_i} \tag{5.4}$$

The necessary condition for a network node to be localized is to have a connection set (I_i) containing more than three elements. If number of elements of the connection set is greater than three, then the unknown position of the ith network device $C_i \equiv (x_{C_i}, y_{C_i}, z_{C_i})$ can be estimated performing an iterative minimization of the following Fitness Function (FF):

$$FF(C_i) = \sum_{j \in I_i} (\tilde{d}_{C_i,R_j} - d_{C_i,R_j})^2 \quad (5.5)$$

being:

- d_{C_i,R_j}, the Euclidean distance between the jth reference point (R_j) and the position of the ith network device (C_i):

$$d_{C_i,R_j} = \sqrt{(x_{C_i} - x_{R_j})^2 + (y_{C_i} - y_{R_j})^2 + (z_{C_i} - z_{R_j})^2} \quad (5.6)$$

- $C_i \equiv (x_{C_i}, y_{C_i}, z_{C_i})$, the position of the ith network device to be localized in the localisation space $\xi \subseteq \mathbb{R}^3$.
- $R_j \equiv (x_{R_j}, y_{R_j}, z_{R_j})$, the position of the jth reference point defined by the multiple repositioning of the reference artefact.

From a computational viewpoint the algorithm is quite simple to be implemented and run. An intuitive drawback is the need for a significant human moderation: every time the reference artefact is moved it has to be located in a global reference coordinate system. To perform this operation the operator has to define a global reference coordinate system and manually locate the position of the reference artefact by means of external devices (laser rules, laser levels, etc.). It is obvious that the accuracy of this operation influences the accuracy of the final localisation of the network devices.

5.5.2 Second Procedure

The goal of the second procedure is that of getting free from the constraint of the a priori localisation of the reference artefact otherwise needed in the first procedure. In this case, the reference artefact is not used for directly locating the network devices, but just to obtain distance information.

This localisation procedure can be divided in two phases:

- *First Phase* The method uses multiple replacements of the dimensional reference artefact to evaluate the mutual distances between the network devices (see Fig. 5.8). The reference artefact is kept still in multiple positioning under the network devices as for the first procedure. Every repositioning of the reference artefact univocally defines a new local coordinate system. All network devices that are able to estimate the distances to three or more embedded devices, are

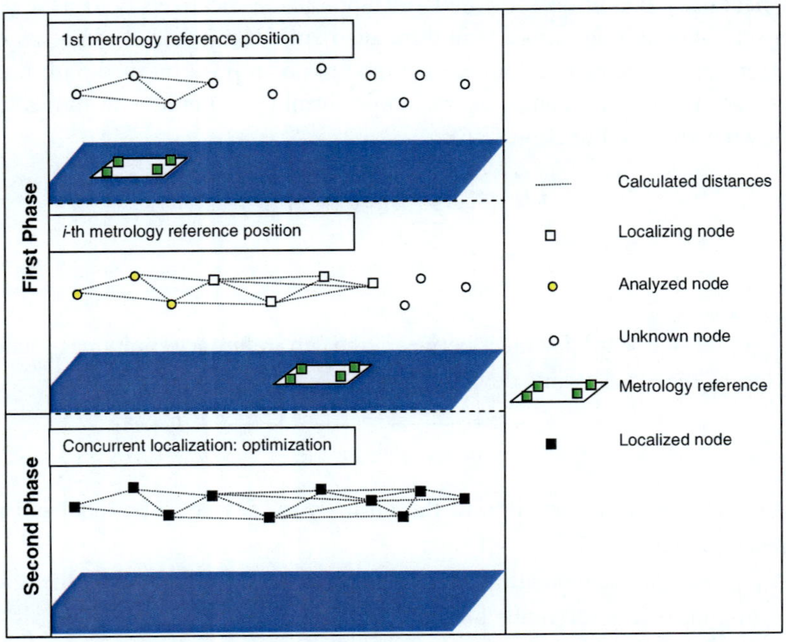

Fig. 5.9 First phase and second phase of the second localisation procedure. Each positioning of the reference artefact defines a set of distances

localized in the local coordinate system by solving a non-linear problem similar to that defined in Eq. 5.5. If two devices are localized in the same local coordinate system then it is possible to calculate their Euclidean distance according to Eq. 5.6.

- *Second Phase* Given the distances between network devices, it is possible to find their position by solving an optimisation problem. The optimisation searches for the global minimum of a fitness function whose goal is to identify the network layout that better satisfies the distance constraints (Fig. 5.9).

Let N be the number of network devices to be localised, and $C_{i,k} = (x_{C_{i,k}}, y_{C_{i,k}}, z_{C_{i,k}})$ and $C_{j,k} = (x_{C_{j,k}}, y_{C_{j,k}}, z_{C_{j,k}})$ the position of two network devices localized in the same local coordinate system defined by the kth reference artefact repositioning. Then let $(\tilde{d}_{C_i,C_j})_k$ be their Euclidean distance calculated as

$$\tilde{d}_{C_{i,k},C_{j,k}} = \sqrt{(x_{C_{i,k}} - x_{C_{j,k}})^2 + (y_{C_{i,k}} - y_{C_{j,k}})^2 + (z_{C_{i,k}} - z_{C_{j,k}})^2} \quad (5.7)$$

Now let $\tilde{d}_{C_i,C_j}$ be the average of the Euclidean distances defined by k different local coordinate systems:

$$\tilde{d}_{C_i,C_j} = \frac{\sum_{k=1}^{K} \tilde{d}_{C_{ik}C_{jk}}}{K} \quad (5.8)$$

where K is the total number of reference artefact repositioning in which both C_i and C_j have been localised.

Collecting all these distances derived from every repositioning of the reference artefact, it is possible to build a global distance matrix which is obviously independent of the different local coordinate systems. The resulting distance matrix is defined as:

$$(d_{i,j}) = \begin{cases} \tilde{d}_{C_i,C_j} & \text{If } C_i \text{ and } C_j \text{ can be localized on the same local coordinate system} \\ 0 & \text{Otherwise} \end{cases} \tag{5.9}$$

As for the first localisation procedure, for each network device it is possible to define its connection set I_i as:

$$I_i = \{j \in \{1, \ldots, M \cdot P\} : \tilde{d}_{C_i,C_j} \neq 0\}_i \tag{5.10}$$

So far, the solution of a global optimisation problem allows the concurrent localisation of the network devices. The Fitness Function to be minimized is similar to that of Eq. 5.5:

$$FF(C_1, \ldots, C_N) = \sum_{i=1}^{N} \sum_{j \in I_i} \left(\tilde{d}_{C_i,C_j} - d_{C_i,C_j}\right)^2 \tag{5.11}$$

being:

- N the number of network devices C_i;
- $C_i \equiv (x_{C_i}, y_{C_i}, z_{C_i})$ the unknown position of point C_i in the localisation space $\xi \subseteq \mathbb{R}^3$;
- d_{C_i,C_j} the Euclidean distance between $C_i \equiv (x_{C_i}, y_{C_i}, z_{C_i})$ and $C_j \equiv (x_{C_j}, y_{C_j}, z_{C_j})$.

The unknown optimisation variables are the three spatial coordinates $(x_{C_i}, y_{C_i}, z_{C_i})$ for $i = 1, \ldots, N$. The optimisation is an iterative procedure: starting from a network layout of first approximation, the algorithm iteratively refines it in order to better satisfy distance constraints. In general the optimisation convergence towards the correct layout is not a priori granted, but it strongly depends on the first approximation solution. In order to avoid such kind of problems a solution proposed and tested on different WSNs has been adopted for the MScMS-I (Priyantha et al. 2005).

While the first localisation procedure starts from the ideal condition in which each reference repositioning is done in a known position, this procedure is free from any constraint about reference artefact positioning. On the other hand, it may be prone to some errors due to particular network layouts with a inhomogeneous node distribution. A necessary condition for the algorithm to work is that, for each node, the distances to at least other four nodes must be known.

5.5.3 Third Procedure

In some conditions, it is possible to overcome the main drawbacks of both the first and the second method. While the first one can be classified as an *anchor-based* algorithm, the second method is an *anchor-free* algorithm. It does not need an a priori localisation of the reference artefact, thus resulting handier and more scalable. On the other hand, it may lack accuracy because it does not consider some of the available information about the local localisations.

The third method is quite similar to the second one, except for a further assumption. In practice, it is often customary to move the reference artefact on a plane, for example on the floor of the working area. This method is based on the assumption of planarity of the surface of the reference artefact support. The local coordinate systems resulting from the reference artefact repositioning have the same z coordinate axis.

The procedure is articulated in two phases:

- *First Phase* This method uses multiple replacements of the reference artefact as for the second localisation procedure. Each time the reference artefact is moved, it defines a local coordinate system. Besides storing mutual distances, during this phase it also saves the z coordinates of each network device.
- *Second Phase* The optimisation searches for the global minimum of a fitness function whose goal is to identify the network layout that better satisfies distances constraints and z measurements.

As for the first and the second procedure, the reference artefact is moved under the network devices. Every repositioning of the reference artefact univocally defines a new local coordinate system. Let denote with $\mathrm{C}_{i,k} \equiv \left(x_{\mathrm{C}_{i,k}}, y_{\mathrm{C}_{i,k}}, z_{\mathrm{C}_{i,k}}\right)$ the local coordinates of the ith network device referred to the kth repositioning of the reference artefact respectively. If, for every repositioning of the reference artefact, the defined local coordinate system has the z-axis orthogonal to the support plane, then what changes between the different local coordinate systems are just x and y coordinates. Let $\tilde{z}_{\mathrm{C}_i}$ be the average of all $z_{\mathrm{C}_{i,k}}$ obtained localizing the same network device (C_i) in different local coordinate systems. The localisation problem can be lead to a bi-dimensional problem.

As for the second method, if two devices can be localized during the same reference artefact repositioning, it is possible to calculate their bi-dimensional Euclidean distance as:

$$\tilde{d}_{\mathrm{C}_{i,k},\mathrm{C}_{j,k}} = \sqrt{\left(x_{\mathrm{C}_{i,k}} - x_{\mathrm{C}_{j,k}}\right)^2 + \left(y_{\mathrm{C}_{i,k}} - y_{\mathrm{C}_{j,k}}\right)^2} \quad (5.12)$$

Let $\tilde{d}_{\mathrm{C}_i,\mathrm{C}_j}$ be the average of all the bi-dimensional distances between C_i and C_j defined by all the reference artefact repositioning.

$$\tilde{d}_{C_i,C_j} = \frac{\sum_{k=1}^{K} \tilde{d}_{C_{i,k},C_{j,k}}}{K} \tag{5.13}$$

where K is the total number of reference artefact repositioning in which both C_i and C_j have been localised. So far it is possible to define a distance matrix **D** as in Eq. 5.9.

The optimisation algorithm of the second phase starts from a raw first approximation bi-dimensional layout (Priyantha et al. 2005) and it searches for the minimum of a fitness function:

$$FF((x_1, y_1)\ldots(x_N, y_N)) = \sum_{i=1}^{N} \left[\left(\sum_{j \in I_i} \left(\tilde{d}_{C_i,C_j} - d_{C_i,C_j} \right)^2 \right) \right] \tag{5.14}$$

where:

- N is the number of devices to be localized;
- (x_i, y_i) are the x and y unknown coordinates of the ith network device in the localisation space $\xi \subseteq \mathbb{R}^3$;
- d_{C_i,C_j} is the bi-dimensional Euclidean distance between C_i and C_j;
- $I_i = \{j \in \{1, \ldots, M \cdot P\} : \tilde{d}_{i,j} \neq 0\}_i$ the connection set of node C_i.

Once the optimisation phase is concluded, the three-dimensional coordinates of each network node are then $C_i \equiv (x_{C_i}, y_{C_i}, \tilde{z}_{C_i})$.

This method has been developed as a compromise between the two previous ones. Assuming one knows that the reference artefact is placed on a flat surface, this method handles the local calculations of the z coordinates as additional measurements. The necessary condition for the algorithm to work is relaxed, compared to that of the second method, as, for each node, the distances to other three nodes must be known.

5.5.4 Tests and Performance Comparison

The three proposed localisation procedures were tested on different network topologies. The experimental trials were run in the Industrial Metrology and Quality laboratory of DISPEA—Politecnico di Torino. During the experimental measurements the temperature was kept constant at about $T = 21°$ with relative humidity $RH = 27\%$. In these conditions the speed of sound value (s) was set to 343 m/s.

The experimental tests were carried out with a number of network devices horizontally placed on the ceiling of the laboratory. In general, this kind of positioning allows a better coverage of the working area and, at the same time, avoids the presence of obstacles between the contact probe and the network devices.

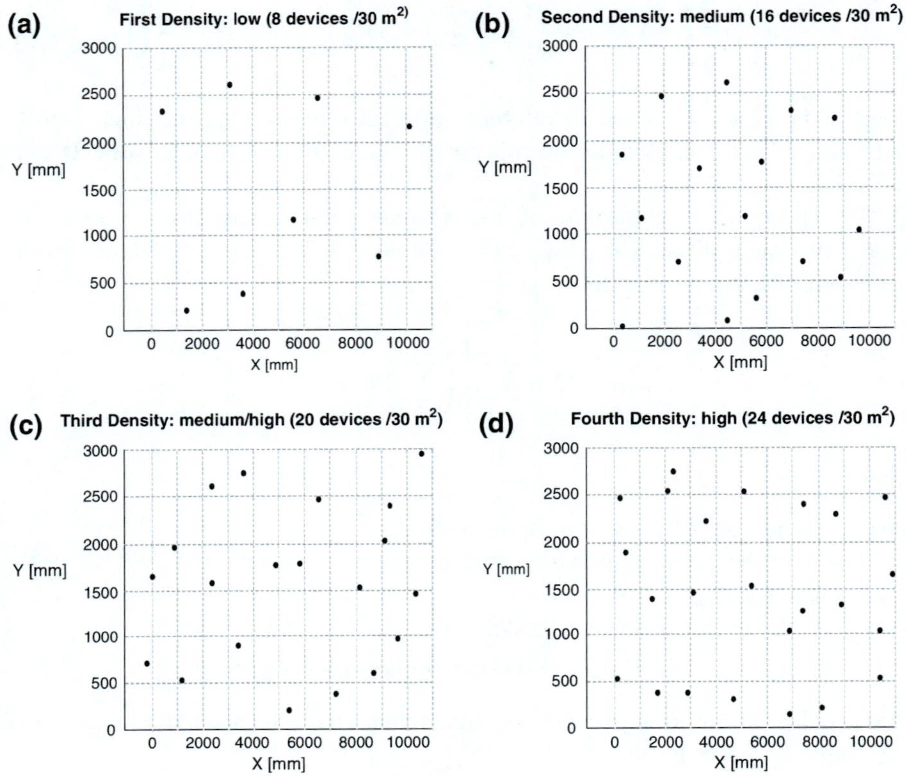

Fig. 5.10 The four network topologies tested in order to evaluate the effect of network topology on localisation

In order to evaluate the sensor network effect on the localisation four different network topologies, fitting the same working volume (about 90 m^3), were considered:

- low density network: 8 devices randomly placed in a 30 m^2 area (see Fig. 5.10a);
- medium density network: 16 devices randomly placed in a 30 m^2 area (see Fig. 5.10b);
- medium/high density network: 20 devices randomly placed in a 30 m^2 area (see Fig. 5.10c);
- high density network: 24 devices randomly placed in a 30 m^2 area (see Fig. 5.10d).

For each of the four network topologies the reference positions plotted in Fig. 5.10 have been measured using a laser tracker, a metrology instrument whose nominal accuracy is at least two orders of magnitude greater than the MScMS-I accuracy (Andersen 2008).

The dimensional reference artefact presented in Fig. 5.7 was used for the tests. Five devices were placed on each vertex and in the middle of a 700 mm side square. The relative positions of the devices embedded in the reference artefact were calculated using a CMM.

Using a laser tracker it was possible to draw a grid of about 1,500 mm of side so as to know the position of the reference artefact for each repositioning. In order to completely cover the working area, the reference artefact was moved in seven different positions.

During each reference artefact positioning, a PC stored the distance estimation between the reference artefact and the network devices, averaging the measurements for 30 s. It has to be noted that the Cricket working frequency is set to 1 Hz, in order to reduce the natural variability of distance measurements among Crickets (Priyantha et al. 2005).

The same data collected by the reference artefact were used off-line as input of the three different procedures. The data collection was replicated five times.

In order to evaluate the performances of the three proposed localisation techniques, the network layouts produced by the algorithms were compared to the reference network layout given by the laser tracker. To do that, the results produced by the algorithms were roto-translated to best fit the network layout given by the laser tracker. This operation, which is just a rigid transformation, is necessary whenever there is a need to compare localisations given in different reference systems. In detail, a Robust Least Squares (RLS) fitting method was used to reduce the influence of outlier points on the fitting results (Andersen 2008).

To compare the nominal and the computed network layouts, the nominal position of each node and the position obtained by the proposed procedures were compared. If $\mathbf{x}_i$ is the position of the ith network node produced by the localisation procedure and $\mathbf{X}_i$ its nominal position, the localisation error ε can be defined as:

$$\varepsilon = \sqrt{\sum_{i=1}^{N} \frac{\|\mathbf{x}_i - \mathbf{X}_i\|^2}{N}} \tag{5.15}$$

where N is the number of localized devices for each network density (Wu et al. 2008; Savvides et al. 2001).

Figure 5.11 summarizes the results obtained from the experimental tests described in the previous sections. The second localisation method showed the worse performance, while the first was the method that provides the best results according to the proposed indicator. As expected, the performance of the third procedure were closer to that of the first method.

In order to complete the analysis, Table 5.1 provides a view of the times needed by each localisation procedure in terms of set-up, acquisition and computation. Although being qualitative, this analysis can be useful in order to understand the required user involvement and the carrying out complexity of each procedure (see the last column of Table 5.1). The evaluation was codified on three levels: High (H), Medium (M), and Low (L).

The set-up time is the time needed for the procedure arrangement. The first procedure necessitates an a priori localisation of all the positions of the reference

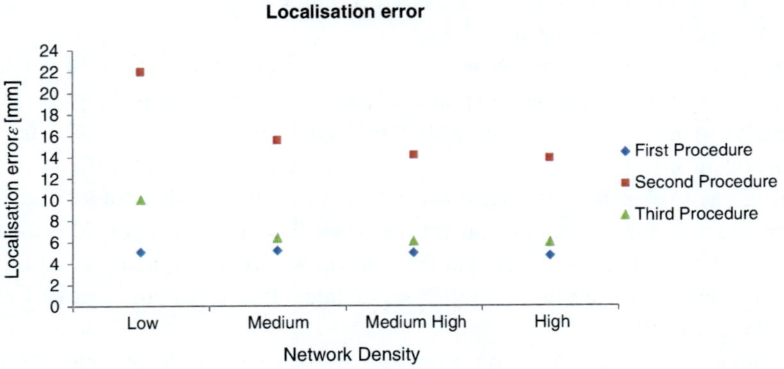

Fig. 5.11 Localisation error according to the error indicator defined in Eq. 5.15

Table 5.1 Time required by each localisation procedure

	Time			Complexity	
	Set-up	Acquisition (min)	Computation (s)	User involvement	Set-up Complexity
First method	2 h	14	< 10	H	H
Second method	1 min	7	< 60	M	L
Third method	1 min	7	< 30	M	L

Data are referred to a medium density network topology. The Complexity has been qualitatively evaluated on three levels: Low (L), Medium (M) and High (H)

artefact and a manual data entry. This operation may take up to some hours depending on the size of the network. The second and the third method do not need any particular a priori setting apart from the software and connections start-up.

The acquisition time is the time taken to position the reference artefact and to perform data acquisition. It is obviously dependent on the number of repositioning (7 in our experiment). For each procedure the reference was kept still for about 30 s in each position. A reference handling time between each acquisition of a further 30 s has been considered for the second and the third procedures, and one and a half minute for the first method that requires an accurate and careful positioning.

The computational time needed by the algorithm to run was small compared to the other times. The personal computer used for this evaluation was an AMD Athlon(tm) 64 Processor 3500+ 2.21 GHz with 1.98 GB RAM.

5.5.5 General Considerations

From the analysis of the results presented in the previous section, it emerges that, above all, the first one is the method that performs better, but it requires a significant user involvement as well as a further metrology instrument able to accurately localize the reference positions. This method may be used in particular

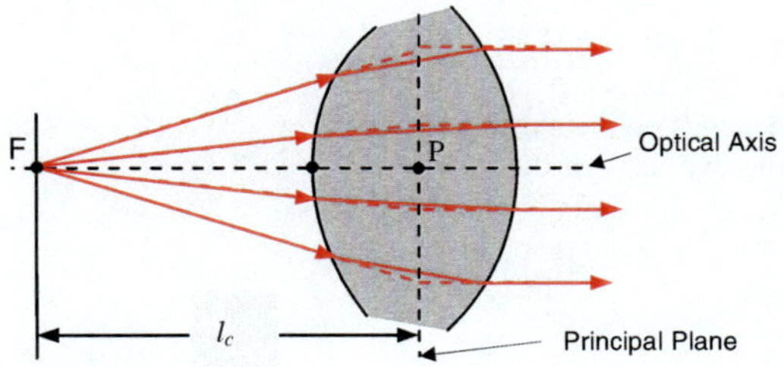

Fig. 5.12 Focal and principal points of a compound lens in air. F is the front focal point, P the principal point. The focal distance l_c is the distance between F and P

contexts in which a large-scale metrology instrument is available and can be used just once to define the reference artefact positions.

The second method, on the other hand, requires modest user involvement, being free from any constraint. The drawback is the low performance level on sensor networks not uniformly distributed, as in the tested experimental case.

Among all the examined procedures, the third method seems to be the best compromise, not needing any particular set-up, but providing a sufficient performance level. On the contrary, this procedure may be inappropriate where the reference support surface is not sufficiently flat.

5.6 MScMS-II Calibration

The multi-camera calibration problem of MScMS-II is faced by using a fully automatic single-point self-calibration technique (Svoboda et al. 2005).

More particularly, the set of unknown parameters ($\chi_{\text{MScMS-II}}$) considered by this calibration procedure is composed by 15 parameters for each camera:

- Nine internal parameters which characterize each camera (Svoboda et al. 2005):
  - Principal point (P), i.e., the point where the principal plane crosses the optical axis (two coordinates on the camera projection plane, see Fig. 5.12);
  - Focal Length (l_c), i.e., the distance from the lens focus (F) and the principal point (see Fig. 5.12);
  - Skew coefficient. It is related to the angle (α_c) between the x and y pixel axes. In case of rectangular pixels α_c is equal to zero (see Fig. 5.13);
  - Distortion coefficients ($\boldsymbol{k}_c$, 5 parameters). These terms are the coefficients of a polynomial that models the image distortion produced by each camera (Fig. 5.14).
- Six external parameters:
  - Position of the camera (x_c, y_c, z_c);
  - Orientation of the camera (ω_c, ϕ_c, κ_c).

This calibration technique requires a minimum of three cameras and a reference artefact for aligning and scaling the reference system (see Fig. 5.15).

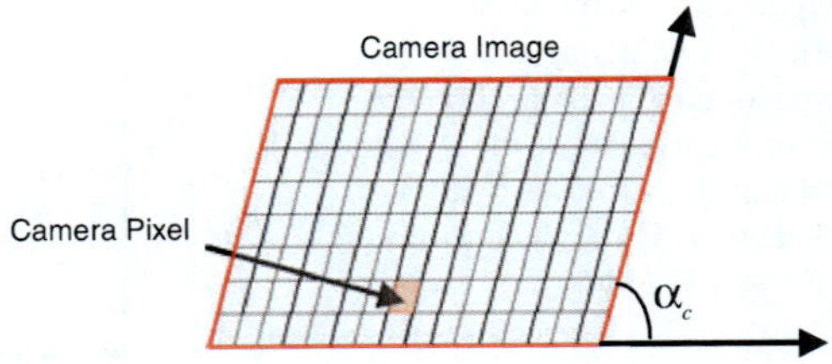

Fig. 5.13 The skew coefficient is a parameter that depends on the angle between the two axes of the pixels. When the angle is equal to zero, the skew coefficient is null

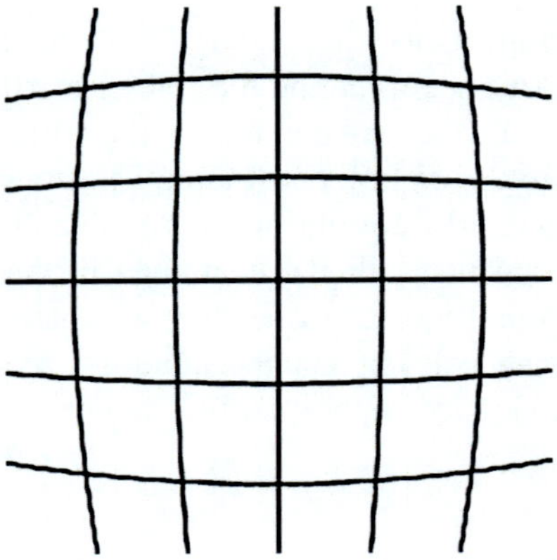

Fig. 5.14 An example of optical radial distortion. Distortion is a deviation from rectilinear projection, a projection in which straight lines in a scene remain straight in an image. The distortion coefficients $\boldsymbol{k_c}$ allow to model and correct this phenomenon

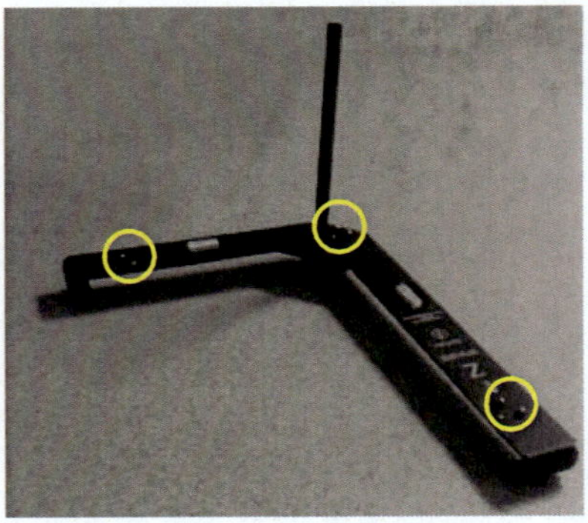

Fig. 5.15 Dimensional reference artefact used for the calibration procedure of the MScMS-II. In this case the number of embedded devices is three. Device housings are highlighted with circles (Galetto et al. 2011). (with permission)

According to the calibration procedure, a single reflective marker is randomly moved in several unknown positions within the working volume and tracked by the IR camera sensors. Image acquisition and processing are managed by the camera onboard hardware that directly provides pixel coordinates in its view plane.

The camera on-board tracking capabilities save the automatic spot detection procedure, which is a computationally expensive operation. On the other hand, the camera self-calibration are affected by the reliability of the embedded tracking engine, i.e., by its capabilities to correctly identify the bright spot in the image and to calculate its position in the camera view plane.

Firstly, the calibration algorithm performs a two-step procedure for discarding outliers, either due to reflections in the working environment or measurement errors of the tracking engine. False points are removed from the list of visible points of the IR cameras according to an iterative pair-wise analysis and a 2D reprojection error-based strategy. As to the former step, image pairs are iteratively selected according to the number of visible corresponding points. Point-to-point correspondence is analyzed according to epipolar geometry constraints (Longuet-Higgins 1981; Hartley and Zisserman 2004) and applying a RANSAC-based technique (Fischler and Bolles 1981) for discarding outliers. Survived points are further evaluated by projecting them back to the camera pairs and applying a threshold method based on 2D reprojection errors for removing false ones.

After the outlier filtering, the calibration algorithm implements an iterative procedure that compute the projective structure (i.e., the projection matrix and the reconstructed 3D points' cloud) until outliers are completely removed and estimate a nonlinear distortion model of the camera lens until a stopping condition (either based on a user-defined threshold or a maximum number of iterations) is reached.

Finally, this procedure yields to the eleven internal and external camera parameters described above (Svoboda et al. 2005). As the external camera parameters are provided in an unknown reference frame, having the origin in the centre of the points' cloud, a further step for aligning and scaling the coordinate system is performed using the reference artefact presented in Fig. 5.15.

5.7 iGPS™ Calibration

Each iGPS™ transmitter is characterized by eleven parameters for 3D position measurement. More particularly, the set of unknown parameters (χ_{iGPS}) considered by the iGPS™ calibration procedure is composed by:

- Five internal parameters:
  - the tilt angles of the two laser beams from the spin axis (ϕ_1 and ϕ_2, see Fig. 5.16);
  - the cone central angles of beam 1 and beam 2, which specify the degree of curvature in each beam (two parameters, see Fig. 5.17);
  - the azimuth angle difference of beam 1 and beam 2 (β, see Fig. 5.16);
- Six external parameters:
  - positions of the rotating heads (x_t, y_t, z_t);
  - orientations of the rotating heads (ϕ_t, θ_t, ρ_t).

Contrary to MScMS-II, the calibration of the internal parameters is typically made separately from the external parameters calibration (Hedges et al. 2003). This is due to the fact each transmitter is univocally assigned to a system and there is no need for multiple calibrations unless in case of particular accidents

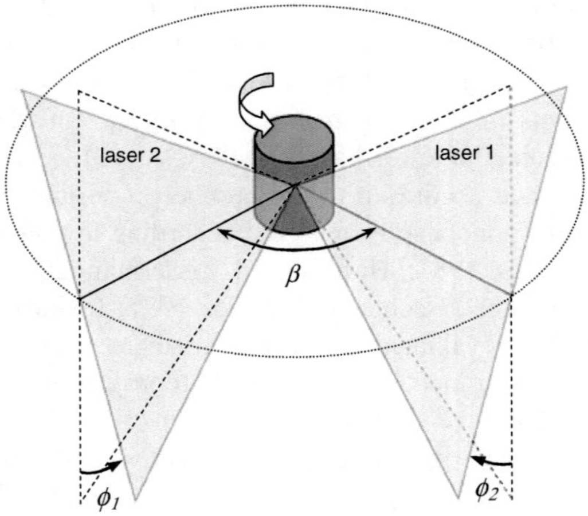

Fig. 5.16 Tilt and azimuth angles of an iGPS™ transmitter

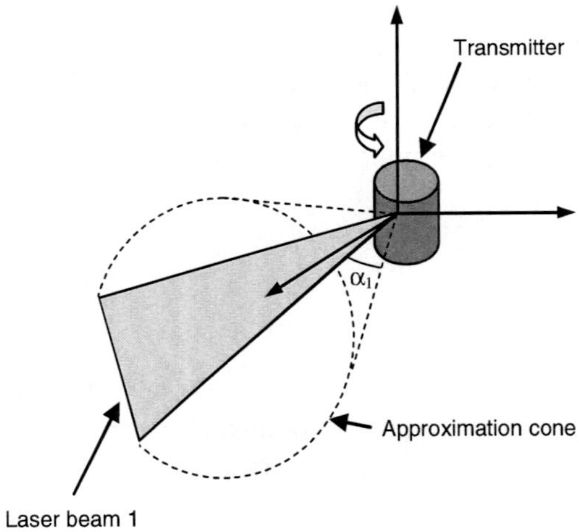

Fig. 5.17 Cone central angle of the first beam. Although it can be considered a plan as a first approximation, the shape of the laser beam is not perfectly planar. Its shape could be rather approximated to a portion of a cone (Hedges et al. 2003). In the schematisation, α_1 is the cone central angles of the first laser beam, which specify its degree of curvature. Since, for each transmitter, there are two laser beams, then two degree of curvature must be in order to fully characterise each device

(e.g., shocks, breakdowns etc.) or maintenance activities. Hedges et al. (2003) describe one possible internal parameters calibration process which is also reported into the patent. In particular the technique used for internal parameter calibration is a resection commonly used in photogrammetry and surveying.

On the other hand, the procedure for the calibration of the external parameters is different. This more general engineering problem is often referred to as "three-dimensional reconstruction" and occurs in areas as diverse as surveying networks

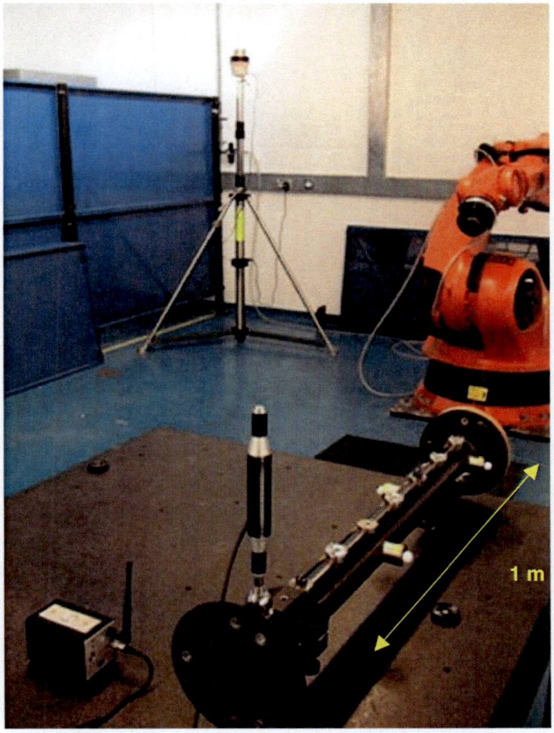

Fig. 5.18 The iGPS™ portable probe measuring a point on a calibrated bar in the Industrial Metrology Laboratories of Bath University. (with permission)

(Wolf and Ghilani 1997), photogrammetry and computer vision (Triggs et al. 2000; Lourakis and Argyros 2009; Ferri et al. 2010).

From the practical point of view, the external parameters calibration is typically carried out by moving the iGPS™ portable probe within the measurement volume. Alternatively a couple of photodiodes mounted on a calibrated and customized bar can be used. The system collects a sufficient number of dynamic measurements to be subsequently processed. An earlier version of the calibration procedure used a collection of static measurements of some points between the transmitters.

As for MScMS-II, before the processing, the collected points are filtered in order to discard outliers due to possible laser beam reflections or noise. Finally the estimation of three-dimensional point coordinates together with transmitter positions and orientations, are obtained through a bundle adjustment procedure (Triggs et al. 2000). This procedure, which is quite common in photogrammetry systems, is aimed at obtaining a network layout reconstruction which is optimal according to a pre-specified fitness function.

The position coordinates of the transmitters with reference to a user defined world coordinate reference system are calculated by defining a scale, which is the absolute distance between two known points (for example the distance between the two photodiodes embedded into the portable probe bar is about 202 mm). Later on, the scaling factor can then be improved by using larger reference lengths later on in the process (Fig. 5.18).

Chapter 6
Self-Diagnostic Tools

6.1 Introduction

Several different causes may affect metrological performance of distributed systems over time: variations of environmental factors, deterioration of some subsystems and other uncontrollable effects (Bosch 1995; Cauchick-Miguel et al. 1996; Franceschini et al. 2002; ISO 10360-2 2001). The option for an on-line evaluation of possible malfunctions is particularly interesting for system users. It would allow a timely correction of anomalies, limiting scraps and low-quality production. Furthermore, keeping under control system metrological characteristics, for ISO 9000 certified companies, is a mandatory requirement (ISO 10012-2 1997).

Metrology systems can be typically subject to three types of verifications (ISO 10360-2 2001):

- initial verification or acceptance test (the acceptance test is normally long, complex and expensive);
- periodic verifications (verifications must be brief, simple to perform and low-cost);
- irregular/occasional controls.

Typical common elements of such verifications are the use of more or less complex and costly artefacts, the use of experienced and qualified personnel, and the need to operate off-line when the machine does not work. Furthermore, all these verifications allow the detection of a possible damage-state, only at the time in which they are carried out. They do not allow establishing the moment when such damage occurred, nor the causes.

Therefore, it is evident the interest for methods that, placing side by side the above verification strategies, are able to automatically display the occurring of any decays in the machine performances. A first result of on-line approaches is the possibility to indicate the need for a more accurate test or even for a complete calibration only when this is really necessary. Second aspect concerns the "guarantee" that the dimensions of the measured part are really those declared by the instrument provider.

F. Franceschini et al., *Distributed Large-Scale Dimensional Metrology*,
DOI: 10.1007/978-0-85729-543-9_6, © Springer-Verlag London Limited 2011

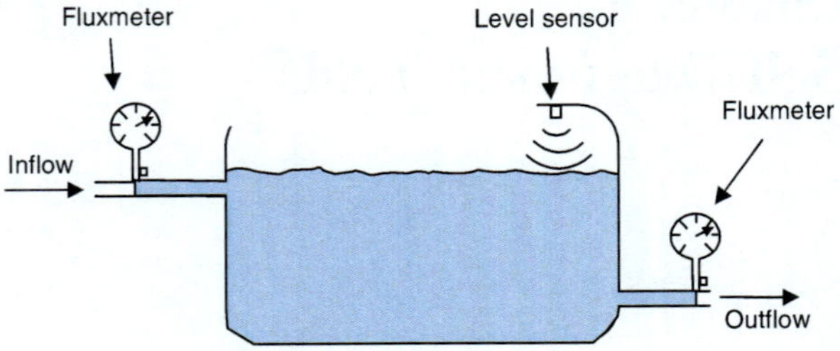

Fig. 6.1 An example of self-diagnostics. The tank presented herein is equipped with two fluxmeters and a level sensor placed on the roof. The two fluxmeters monitor respectively the inflow and the outflow. If the difference between the two flows does not match the level of the tank, this indicates the presence of some leakage

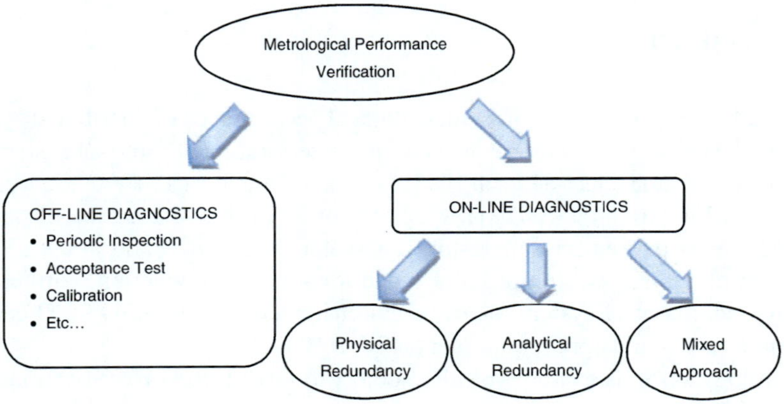

Fig. 6.2 Summary of different approaches for metrological performance verification (Franceschini et al. 2002). (with permission)

The present chapter introduces first the concepts of "self diagnostics" and "reliability of a measurement". Second it provides an analytical description of three on-line diagnostic tests aimed at verifying the metrological performance of distributed systems during the measurement phase. It has to be said that these diagnostics complement the set of diagnostics that usually equip metrology instruments.

6.2 Self-Diagnostics

The problem of "self-diagnostics" in automated systems is not a recent matter, and traditionally many strategies have been proposed (Clarcke 1995; Henry and Clarcke 1992; Isermann 1984). Self-diagnostics can be defined as the capability of a system to identify and signal possible anomalies which can affect its proper functioning. As an example of this concept, let us consider the system presented in

Figs. 6.1. Such a system is able to autonomously identify eventual tank leakages by comparing the inflow and the outflow with the tank level measured by an ad hoc sensor.

In many critical sectors, such as the aeronautical and nuclear ones, the most used techniques for self diagnostics are based on the replication of the instrumentation and the control equipments. This approach is usually very expensive.

In general, it is possible to distinguish two main types of redundancy (see Fig. 6.2):

- The "physical redundancy" (Gertler 1998), which consists of replicating the instrumentation and the control equipments (for example, by integrating the touch probe with an optical device or by creating a redundancy of the probe itself).
- The "analytical" or "model-based redundancy" (Frank 1990; Isermann 1984, 1993), which substitutes the replication of a physical instrumentation by the use of appropriate mathematical models. These latter may derive from physical laws applied to experimental data or from self-learning method (for example, neural networks).

Often the two approaches are employed together. In this case the analytical redundancy method is enhanced with the use of an external "witness-part" (Franceschini et al. 2002). The diagnostic tests proposed into this chapter are based on analytical redundancy.

6.3 The Concept of Measurement Reliability

If we refer to the field of dimensional metrology instruments and CMMs, in particular, the concept of "on-line metrological performance verification" is strictly related to the notion of "on-line self-diagnostics " (Gertler 1998; Franceschini and Galetto 2007). In a sense, this approach is "complementary" to that of uncertainty evaluation presented in Chap. 8 (ISO/TS 15530–6 2000; Phillips et al. 2001; Savio et al. 2002; Piratelli-Filho and Di Giacomo 2003; Feng et al. 2007).

In general, the on-line measurement verification is a guarantee for the preservation of measurement system characteristics (including accuracy, repeatability, and reproducibility) (JCGM 100:2008 2008). The effect of measuring system degradation is the production of "non-reliable measurements".

In general, we can define the concept of "reliability of a measurement" as follows. For each measurable value x, we can define an acceptance interval [LAL, UAL] (where LAL stands for Lower Acceptance Limit and UAL for Upper Acceptance Limit): $LAL \leq x \leq UAL$.

The measure y of the quantity x, obtained by a given measurement system, may be considered as the realization of a random variable Y. It is considered "reliable" if $LAL \leq y \leq UAL$.

Therefore, the I and II type probability errors (misclassification rates) respectively correspond to:

$$\alpha = \Pr\{y \notin [LAL, UAL]\,|LAL \leq x \leq UAL\} \tag{6.1}$$

and

$$\beta = \Pr\{LAL \leq y \leq UAL|x \notin [LAL, UAL]\} \tag{6.2}$$

Usually *LAL* and *UAL* are not a priori known, but are defined according to the metrological characteristics of the measurement system (accuracy, reproducibility, repeatability, etc.), as well as the required quality level of the measurement result (JCGM 100:2008 2008).

The basic principle of the self-diagnostic tools described in this chapter is to define an acceptance interval. If the measurement value (y) is included in this interval, the acceptance test gives a positive response and the measured result is considered reliable. Otherwise, the measurement is rejected (Franceschini et al. 2009b).

6.4 Distance Model-Based Diagnostics

This section presents the first self-diagnostic test. This test is based on the a priori knowledge of the geometry of the mobile probe of the distributed metrology system. The mobile probe generally consists of a rod embedding two active devices or passive targets (A and B) that have to be localised by the systems.

The distance between the two probe devices ($d_{A,B}$) is an a priori known design parameter. On the other hand, the 3D position of the two probe devices/targets and can be measured by each system. Thus their Euclidean distance can be easily calculated as follows:

$$\tilde{d}_{A,B} = \|x_B - x_A\| = \sqrt{(x_A - x_B)^2 + (y_A - y_B)^2 + (z_A - z_B)^2} \tag{6.3}$$

By comparing the Euclidean distance with the nominal distance, it is possible to define a random variable (residual):

$$\varepsilon_{A,B} \equiv \tilde{d}_{A,B} - d_{A,B} \tag{6.4}$$

If the system is properly working, it is reasonable to associate these residuals to a zero-mean normal distribution with a certain standard deviation $\sigma_{\mathrm{A,B}}$:

$$\varepsilon_{A,B} \sim N(0, \sigma_{A,B}) \tag{6.5}$$

Assuming a risk α as a type I error, a simple statistical test can be performed in order to evaluate the measurement reliability.

Let Q_{MIN} and Q_{MAX} be the $\frac{\alpha}{2}$-quantile and $\left(1 - \frac{\alpha}{2}\right)$-quantile, respectively, of a normal distribution with mean $\mu_{A,B} = 0$ and standard deviation $\sigma_{A,B}$. For a given value of α, Q_{MIN} and Q_{MAX} can be expressed as multiples of the standard deviation σ_{AB}:

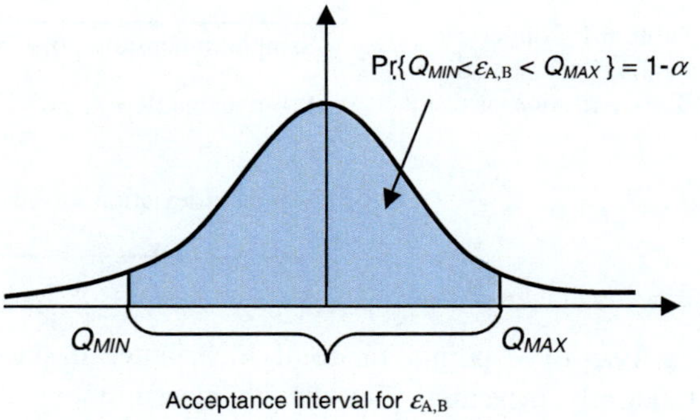

Fig. 6.3 Definition of the acceptance interval for residuals. If a generic measurement produces a residual value outside the acceptance interval then the measurement is considered unreliable

$$\begin{aligned} Q_{MIN} &= z_{\frac{\alpha}{2}} \cdot \sigma_{A,B} \\ Q_{MAX} &= z_{1-\frac{\alpha}{2}} \cdot \sigma_{A,B} \end{aligned} \tag{6.6}$$

where $z_{\frac{\alpha}{2}}$ and $z_{1-\frac{\alpha}{2}}$ are the values of the standardised normal distribution corresponding to $\frac{\alpha}{2}$ and $\left(1-\frac{\alpha}{2}\right)$ levels of probability, respectively. Furthermore σ_{AB} is the standard deviation associated to the natural variability of the system, which can be a priori estimated, during the preliminary phase of set-up and calibration.

Every time a measurement is performed, the experimental residual $\varepsilon^*_{A,B}$ is calculated:

$$\varepsilon^*_{A,B} = \tilde{d}^*_{A,B} - d_{A,B} \tag{6.7}$$

The interval $[Q_{MIN}, Q_{MAX}]$ is assumed as the acceptance interval for the reliability test of the measurement (see Fig. 6.3).

If the calculated residual $\varepsilon^*_{A,B}$ satisfies the condition

$$Q_{MIN} \le \varepsilon^*_{A,B} \le Q_{MAX} \tag{6.8}$$

the measurement cannot be considered unreliable, hence it is not rejected.

6.4.1 Setup of Parameters

The risk α is defined by the user according to the required level of system performance. A high value of α prevents against non-acceptable solutions of the optimisation problem, minimising the type II error β. On the other hand, a low value of α speeds up the measurement procedure, although it might drive to collect wrong data due to the high level of type II error β.

The standard deviation $\sigma_{\mathrm{A,B}}$ can be evaluated by applying the uncertainty composition law, or empirically, on the basis of experimental distance measurements. Here follows an example of procedure for its empirical estimation.

Table 6.1 Numerical details of the statistical characterisation of residuals

Sample dimension: $N_{TOT} = M$	147
Mean estimate: $\hat{\mu}_{A,B} = \sum_{j=1}^{M} \frac{(\varepsilon_{A,B})_j}{M}$	−0.5 mm
Standard deviation estimate: $\hat{\sigma}_{A,B} = \sqrt{\sum_{j=1}^{M} \frac{(\varepsilon_{A,B})_j^2}{M}}$	17.3 mm

A set of M points, randomly distributed in the measurement space $\xi \subseteq \mathbb{R}^3$, are randomly measured. For the jth measurement, the residual $(\varepsilon_{A,B})_j$ is calculated (where $j = 1, \ldots, M$).

In the absence of systematic errors and time or spatial/directional effects, we hypothesize the same normal distribution for all the random variables $(\varepsilon_{A,B})_j$, i.e.:

$$\left(\varepsilon_{A,B}\right)_j \sim N\left(0, \sigma_{A,B}^2\right) \tag{6.9}$$

The standard deviation may be estimated as follows:

$$\hat{\sigma}_{A,B} = \sqrt{\sum_{j=1}^{M} \frac{\left(\left(\varepsilon_{A,B}\right)_j - 0\right)^2}{M}} = \sqrt{\sum_{j=1}^{M} \frac{\left(\varepsilon_{A,B}\right)_j^2}{M}} \tag{6.10}$$

The obtained value of $\hat{\sigma}_{A,B}$ is considered as the reference value for the test. Test limits defined in Eq. 6.6 become:

$$\begin{aligned} Q_{MIN} &= z_{\frac{\alpha}{2}} \cdot \hat{\sigma}_{A,B} \\ Q_{MAX} &= z_{1-\frac{\alpha}{2}} \cdot \hat{\sigma}_{A,B} \end{aligned} \tag{6.11}$$

6.4.2 A Practical Case Study

An empirical investigation was carried out for MScMS-I. In order to estimate $\sigma_{A,B}$ the steps here below have been followed:

- A sample of $M = 147$ points, which were randomly measured by the probe, was considered.
- Coordinates of each probe device were evaluated using the "mass-spring" localisation algorithm (see Sect. 6.5). A sample of 147 residuals was obtained.
- Normal distribution of residuals was tested using a chi-square test (Montgomery 2005).
- Standard deviation of residuals was estimated using Eq. 6.10. The obtained result is $\hat{\sigma}_{A,B} = 17.3$ mm (see Table 6.1 for data details).

The resulting 95% confidence interval for $\varepsilon_{A,B}$ is [−34.0; 34.0] mm. Thus a generic measurement point cannot be considered unreliable if

$$\left|\varepsilon^*_{A,B}\right| \leq 34.0\,\mathrm{mm} \tag{6.12}$$

If, for example, a generic measurement gives a value $\varepsilon^*_{A,B}$ = 38 mm than it is considered unreliable.

6.5 Energy Model-Based Diagnostics

The Energy model-based diagnostics is here specifically developed for MScMS-I although it could easily be extended to the other distributed metrology systems. Whenever MScMS-I produces a localisation result, this test verifies its reliability by analyzing the residuals produced by the localisation algorithm.

More in detail, let's consider the Energy Function ($EF(\boldsymbol{x_P})$) defined in Eq. 6.13. The unknown variable x_P can be estimated by performing the iterative minimization of:

$$EF(\boldsymbol{x_P}) = \frac{\sum_{i=1}^{N} \left(\boldsymbol{d_{C_i}} - \boldsymbol{d_{M_i}}\right)^2}{\boldsymbol{N}} \tag{6.13}$$

being:

- $\boldsymbol{x_P} \equiv (x_P, y_P, z_P)$ the unknown coordinates of device P in the localisation space $\xi \subseteq \mathbb{R}^3$;
- N the number of network devices whose 3D position ($\boldsymbol{x_i} = (x_i, y_i, z_i)$, $i = 1, \ldots, N$) is a priori known within the range of communication of device P;
- d_{M_i}, the measured distance between the ith reference point and P;
- d_{C_i}, the calculated Euclidean distance between the ith reference point and P:

$$d_{C_i} = \sqrt{(x_P - x_i)^2 + (y_P - y_i)^2 + (z_P - z_i)^2} \tag{6.14}$$

The problem of finding a minimum of the function $EF(x_P)$ can be handled as the problem of finding the point of equilibrium for a mass-spring system (lowest potential energy) (Franceschini et al. 2009b).

As an example, let consider the 2D case described in Fig. 6.4. A unitary mass is associated to each network node. The node (x_P) with an unknown location is connected to three reference nodes by three springs. Each of these springs has a rest length equal to the measured distance and a unitary force constant.

Knowing the rest lengths (d_{M_i}) and the coordinates of the three reference points, the system potential energy U is given by:

$$U(x_P) = \sum_{i=1}^{N} \frac{1}{2}\left(\sqrt{(x_P - x_i)^2 + (y_P - y_i)^2} - d_{M_i}\right)^2 \tag{6.15}$$

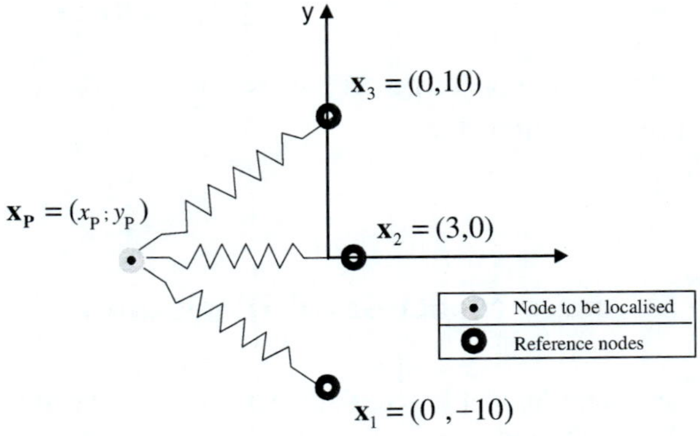

Fig. 6.4 An example of 2D mass-spring system. Three reference nodes (x_1, x_2, x_3) with known location are linked by springs to the point to be localized (x_P)

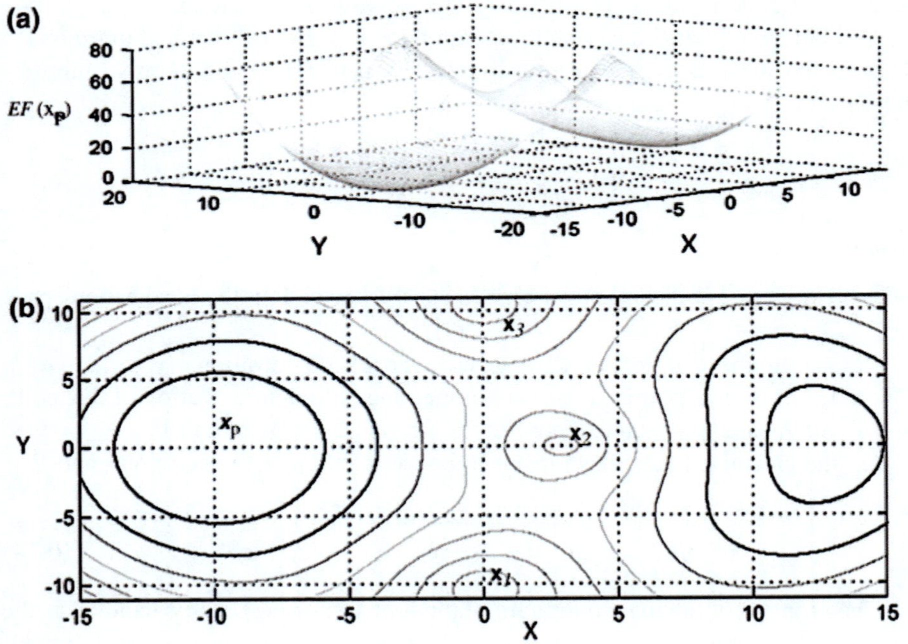

Fig. 6.5 **a** $EF(\boldsymbol{x_P})$ behaviour for the mass-spring system described in Fig. 6.4. Finding the point of minimum means to localize the node P with unknown location. **b** Iso-energetic curves for the mass-spring system described in Fig. 6.3. *Dark grey curves* refer to low energy level, *light grey curves* refer to high energy level. Let notice that $\boldsymbol{x_P}$ is the global minimum point of potential energy. Maxima are in correspondence of the reference points (x_1, x_2, x_3)

Figure 6.5a and b show a 3D and 2D visualization of $EF\ (\boldsymbol{x_P})$ respectively. Since $EF(\boldsymbol{x_P})$is proportional to $U(\boldsymbol{x_P})$, they have the same minima. As expected, the global minimum is where the node to be located actually is, i.e., $\boldsymbol{x_P} \equiv (-10, 0)$.

From what stated above it is clear how $EF(\boldsymbol{x_P}) = 0$ when $d_{M_i} = d_{C_i}$, for $i = 1, \ldots, N$, i.e., when measured distances coincide with calculated ones.

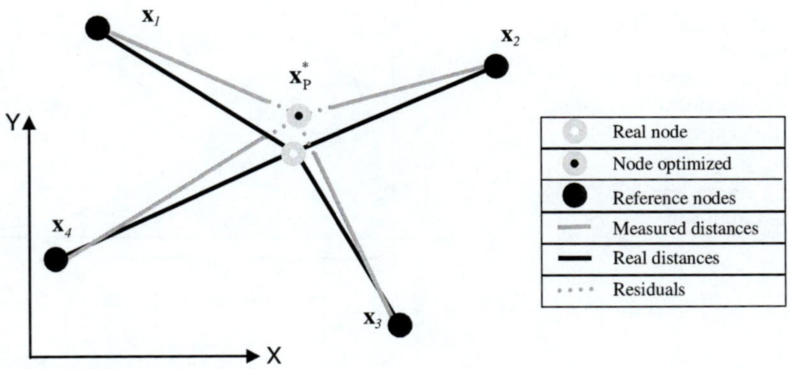

Fig. 6.6 An example of node localisation problem. Measured distances are not equal to actual distances

Because of natural variability of the measuring instrument, two typical situations may occur:

EF ($\boldsymbol{x_P}$) is strictly positive even in the point of correct localisation. This happens because measurements differ, in general, from nominal values;

EF($\boldsymbol{x_P}$) shows a global minimum in a point that is not the correct one. This may occur, for the same reasons, when local and global minima are close to each other (such in the case described in Fig. 6.5). In such a situation, due to the "noise" in distance measurements, a local minimum may turn into a global minimum and vice versa.

The energy model-based diagnostics introduces a criterion to identify the non-acceptable minimum solutions for *EF*($\boldsymbol{x_P}$), in order to prevent system failures. Such criterion enables MScMS-I system to distinguish between reliable and unreliable measurements.

Consider a solution x_P^* to the problem $\min_{x \in \xi} EF(\boldsymbol{x_P})$. In general, if the problem is over determined (i.e., more than three distance constraints in the 3D case and more than two for the 2D case) and the single measurements are affected by noise, the solution satisfying all distance constraints at the same time does not exactly fit the actual node location (see Fig. 6.6).

In such a case, the differences between measured and Euclidean distances may be defined as residuals $\varepsilon_i \equiv (d_{M_i} - d_{C_i})$. Generally, in absence of systematic error causes, it is reasonable to hypothesize a zero mean normal distribution for the random variables ε_i, i.e.:

$$\varepsilon_i = (d_{M_i} - d_{C_i}) \sim N(0, \sigma_i^2) \qquad (6.16)$$

In absence of spatial/directional effects the residual are characterised by the same standard deviation for each node ($\sigma_i^2 = \sigma^2, \forall i$), Eq. 6.13 becomes:

$$EF(\boldsymbol{x_P}) = \sum_{i=1}^{N} \frac{\varepsilon_i^2}{N} = \frac{\sigma^2}{N} \cdot \sum_{i=1}^{N} \frac{\varepsilon_i^2}{\sigma^2} = \frac{\sigma^2}{N} \cdot \sum_{i=1}^{N} \left(\frac{\varepsilon_i}{\sigma}\right)^2 = \frac{\sigma^2}{N} \cdot \sum_{i=1}^{N} Z_i^2 \qquad (6.17)$$

Fig. 6.7 Definition of the acceptance interval for experimental chi-square. If a generic measurement produces an experimental chi-square greater than $\chi^2_{v,\alpha}$, then the measurement is considered unreliable

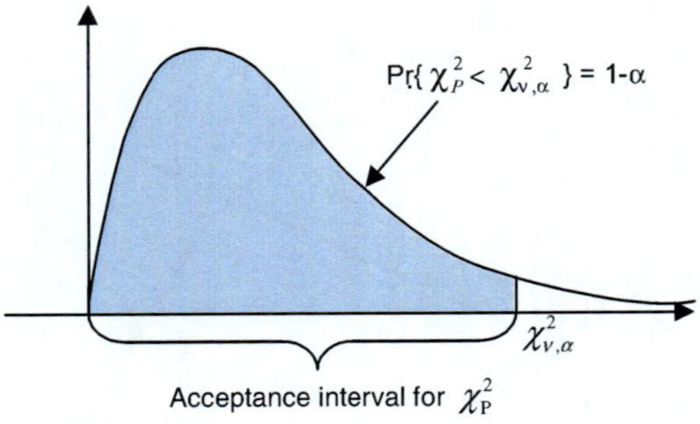

Equation 6.17 can be seen as the sum of N independent normal squared random variables Z_i with zero mean and unit variance, multiplied by the constant term $\frac{\sigma^2}{N}$.

It must be noted that the sum in Eq. 6.17 has only $N - 1$ independent terms. Equation 6.17 causes the loss of a degree of freedom. This implies that, once $N - 1$ terms are known, the Nth one is univocally determined.

Defining a variable $\chi_P^2 = \sum_{i=1}^{N} \left(\frac{\varepsilon_i}{\sigma}\right)^2$, $EF(x_P)$ in Eq. 6.17 has a chi-square distribution with $(N - 1)$ degrees of freedom:

$$EF(\boldsymbol{x_P}) = \frac{\sigma^2}{N} \cdot \chi_P^2 \tag{6.18}$$

The residual variance σ^2 can be a priori estimated for the whole measuring volume, for example during the phase of set-up and calibration of the system.

Every time a measurement is performed for each probe device, the system diagnostics computes the following quantity (experimental chi-square):

$$\chi_P^{2*} = EF(\boldsymbol{x}_P^*)\frac{N}{\sigma^2} \tag{6.19}$$

Assuming a risk α as a type I error, a one-sided confidence interval for variable $\chi^2_{v,\alpha}$ can be calculated. $\chi^2_{v,\alpha}$ is a chi-square distribution with $v = N - 1$ degrees of freedom and $(1 - \alpha)$ confidence level. The confidence interval is assumed as the acceptance interval for the reliability test of the measurement (see Fig. 6.7).

The test drives to the following two alternative conclusions:

- $\chi_P^{2*} \leq \chi^2_{v,\alpha}$ → the measurement is considered reliable, hence it is not rejected;
- $\chi_P^{2*} > \chi^2_{v,\alpha}$ → the measurement is considered unreliable, hence it is rejected and the operator is asked to perform another one.

6.5.1 Setup of Parameters

As usual, the risk level α is established by the user depending on the required level of performance of the system. On the other hand, the estimation of the residual variance can be evaluated in two ways:

- by applying the uncertainty composition law to the calculation of the unknown coordinates (see Chap. 8), starting from the measurement uncertainty of the distances between network devices and probe devices (JCGM 100:2008 2008);
- empirically, on the basis of experimental distance measurements. In this case, the variance of residuals is estimated from a sample of residuals, obtained by measuring a set of points randomly distributed in the whole working volume. This method requires the a priori knowledge of the measured points location. It can be easily implemented during the initial phase of set-up and calibration of the system.

In the following, we focus on this second estimation procedure. Given a set of M points distributed in the measurement space $\xi \subseteq \mathbb{R}^3$, randomly measured by a single Cricket (i.e., with a random sequence of measurement and a random positioning and orientation of the Cricket), for each point j a set of N_j residuals can be calculated, $j = 1, \ldots, M$.

It must be highlighted that the number of residuals N_j may change due to the different number of distances, detected during each measurement.

In the absence of systematic errors and time or spatial/directional effects, it is reasonable to hypothesize the same normal distribution for all the random variables $\varepsilon_{i,j}$, $j = 1, \ldots, M$, $i = 1, \ldots, N_j$, i.e.:

$$\varepsilon_{i,j} = (d_{M_i} - d_{C_i})_j \sim N(0, \sigma^2) \tag{6.20}$$

The residual variance σ^2 may be estimated as follows:

$$\hat{\sigma}^2 = \sum_{j=1}^{M} \sum_{i=1}^{N_j} \frac{(\varepsilon_{i,j} - 0)^2}{\sum_{j=1}^{M} N_j} = \sum_{j=1}^{M} \sum_{i=1}^{N_j} \frac{(\varepsilon_{i,j})^2}{\sum_{j=1}^{M} N_j} \tag{6.21}$$

The obtained value $\hat{\sigma}^2$ is considered as the reference value for the test. With this notation, Eq. 6.19 becomes:

$$\chi_P^{2*} = EF(\boldsymbol{x}_P^*) \cdot \frac{N}{\sigma^2} \cong EF(\boldsymbol{x}_P^*) \cdot \frac{N}{\hat{\sigma}^2} \tag{6.22}$$

6.5.2 A Practical Case Study

An empirical exploratory investigation was carried out to verify the goodness of this approach.

For a network consisting of five reference devices, placed in the measurement volume as schematized in Fig. 6.8, σ^2 has been empirically estimated as follows:

- $M = 253$ randomly distributed points in the working volume were measured by a single Cricket.

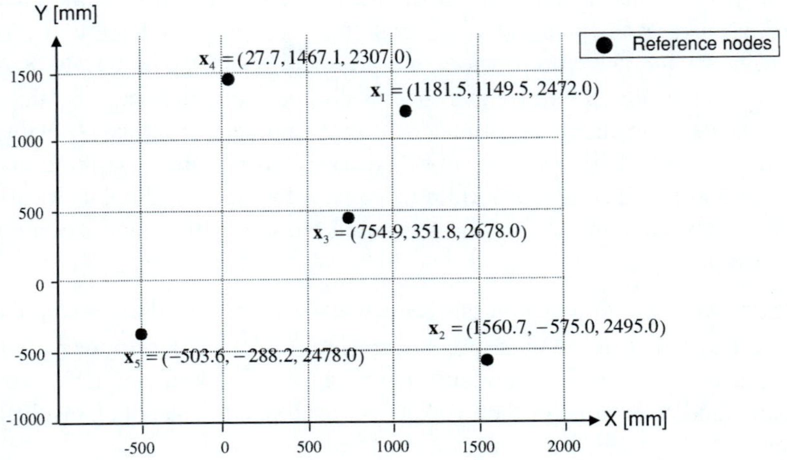

Fig. 6.8 Scheme of the reference node disposal (constellation beacons) in the measurement volume. The coordinates of each device are shown in *brackets*

Table 6.2 Details of data analysis for standard deviation estimation of residuals

Sample dimension: $N_{TOT} = \sum_{j=1}^{M} N_j$	1,123
Mean estimate: $\hat{\mu} = \sum_{j=1}^{M}\sum_{i=1}^{N_j} \frac{\varepsilon_{i,j}}{\sum_{j=1}^{M} N_j}$	0.3 mm
Variance estimate: $\hat{\sigma}^2 = \sum_{j=1}^{M}\sum_{i=1}^{N_j} \frac{(\varepsilon_{i,j})^2}{\sum_{j=1}^{M} N_j}$	100.0 mm^2

- The coordinates $x_j, j = 1, \ldots, M$, of each node were evaluated using the "mass-spring" localisation algorithm and a sample of 1,265 residuals was obtained.
- The normal distribution of residuals was tested using a chi-square test (Montgomery 2005).
- Residual variance was estimated according to Eq. 6.21. The obtained value was $\hat{\sigma}^2 = 100.0\ \text{mm}^2$ (see Table 6.2 for details).

The acceptance limit for $EF(\boldsymbol{x_P})$, assuming a type I error risk $\alpha = 0.05$ and $v = N - 1 = 5 - 1 = 4$ degrees of freedom, is:

$$EF\left(\boldsymbol{x}_{\boldsymbol{P}}^{*}\right) \leq \frac{\hat{\sigma}^2}{N} \cdot \chi^2_{v=4,\alpha=0.05} \Rightarrow EF\left(\boldsymbol{x}_{\boldsymbol{P}}^{*}\right) \leq 189\,\text{mm}^2 \qquad (6.23)$$

Let us consider a typical situation that can occur using ultrasound technology to estimate distances: US reflection. Referring to the configuration in Fig. 6.9, suppose that a generic point $\boldsymbol{x_P}$, inside the measurement volume has to be localised (for example, $\boldsymbol{x_P} \equiv (1067.2, -122.5, 925.8)$).

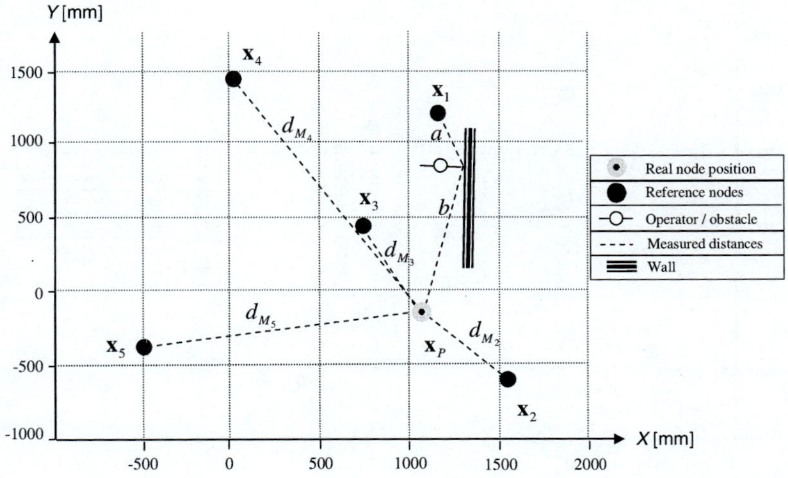

Fig. 6.9 Schematization of a potential misleading situation: walls and obstacles can cause wrong distance estimation. In this case the measured distance between node 1 and node $\boldsymbol{x_P}$ results higher than the actual distance ($d_{M_1} = a + b$)

A device positioned in $\boldsymbol{x_P}$ is able to correctly measure distances from all the network nodes except for one of them. An obstacle (for example, the operator carrying out the measurement) is interposed between $\boldsymbol{x_P}$ and that node, preventing direct US signal propagation. At the same time, a wall placed close to the two devices causes US signal reflection. The consequence is that the pair wise distance estimation between those two devices results 100 mm larger (see Fig. 6.9).

The measured distances are:

$$\begin{aligned} d_{M_1} &= 2104.8\,\text{mm} \\ d_{M_2} &= 1713.4\,\text{mm} \\ d_{M_3} &= 1831.4\,\text{mm} \\ d_{M_4} &= 2355.6\,\text{mm} \\ d_{M_5} &= 2215.2\,\text{mm} \end{aligned} \tag{6.24}$$

In this case, the algorithm produces the following wrong localisation solution (see Fig. 6.10): $\boldsymbol{x}^*_{\boldsymbol{P'}} \equiv (1022.6; -187.3; 911.8)$. According to these results the calculated distances are:

$$\begin{aligned} d_{C_1} &= 2060.7 \text{ mm} \\ d_{C_2} &= 1716.5 \text{ mm} \\ d_{C_3} &= 1865.9 \text{ mm} \\ d_{C_4} &= 2381.9 \text{ mm} \\ d_{C_5} &= 2189.2 \text{ mm} \end{aligned} \tag{6.25}$$

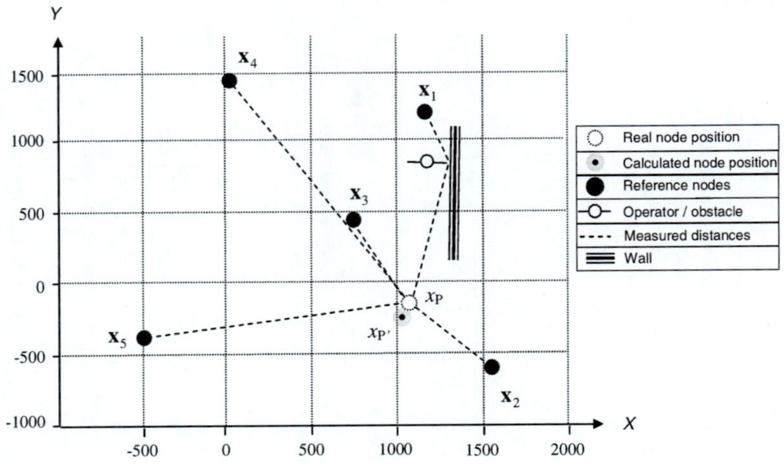

Fig. 6.10 Schematization of a wrong localisation solution (P') due to a wrong distance estimations between node 1 and node P (point coordinates in millimetres [mm]—see Fig. 6.8)

Thus the energy function can be evaluated as:

$$EF\left(\boldsymbol{x}_P^*\right) = \frac{(2104.8 - 2060.7)^2 + \cdots + (2215.2 - 2189.2)^2}{5} \cong 904\,\text{mm}^2 \quad (6.26)$$

It can be concluded that the localisation solution is characterized by a level of "energy" higher than the acceptance limit (see Eq. 6.23): $EF\left(\boldsymbol{x}_P^*\right) \cong 904\,\text{mm}^2 > 189\,\text{mm}^2$. Owing to this result, the energy model-based diagnostics suggests rejecting the measurement.

Removing the obstacle, the distance between node 1 and node P becomes $d_{M_1} = 2004.8\,\text{mm}$, and the correct localisation solution can be obtained:

$$\boldsymbol{x_P}^* \equiv (1067.2; -122.5; 925.8) \quad (6.27)$$

The new "energy" value is: $EF(\boldsymbol{x}_P^*) \cong 41\ \text{mm}^2 < 189\ \text{mm}^2$. Consequently $\boldsymbol{x}_P^*$ cannot be considered unreliable and the measurement is not rejected.

6.6 Sensor Physical Diagnostics

This section deals with the last diagnostic test presented in this chapter. This test is herein specifically developed for MScMS-I. Cricket devices are provided with two ultrasound sensors: a transmitter and a receiver. This allows each device to compute two different distances, denoted as d_T and d_R, referring to transmitted and received US signals respectively.

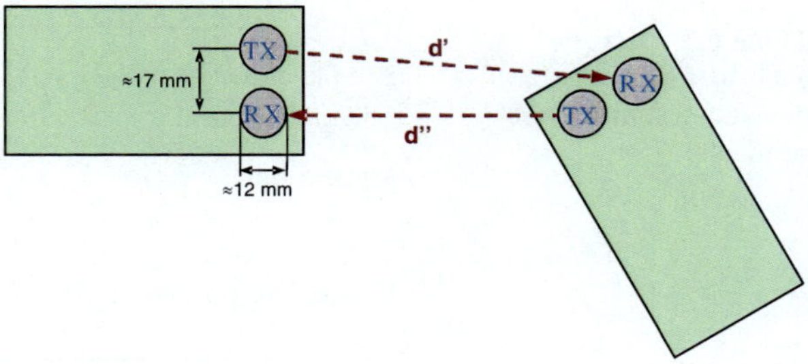

Fig. 6.11 Schematic representation of the difference between "received" and "transmitted" distances for two Crickets

Figure 6.11 shows how the two measured distances can assume different values depending on the orientation of devices. Distances estimated according to the received US signal are used as a further redundancy.

In this case it is possible to study the distribution of the difference between these two measures of distance in order to establish a diagnostic criterion. The random variable ε_{TR} is defined as residual:

$$\varepsilon_{T,R} \equiv d_T - d_R \tag{6.28}$$

As a first approximation it can be assumed a zero mean normal distribution, i.e., $\varepsilon_{T,R} \sim N(0, \sigma_{T,R})$.

Assuming a risk α as a type I error, a statistical test can be performed in order to evaluate measurement reliability.

Acceptance interval limits are defined in the same way as for the distance model-based diagnostics (see Eq. 6.6):

$$\begin{aligned} Q_{MIN} &= z_{\frac{\alpha}{2}} \cdot \sigma_{T,R} \\ Q_{MAX} &= z_{1-\frac{\alpha}{2}} \cdot \sigma_{T,R} \end{aligned} \tag{6.29}$$

Since MScMS-I requires the simultaneous localisation of devices A and B, at least eight distances are evaluated at the same time. Four distances are necessary to locate a point in a 3D space. If the calculated residual $\varepsilon_{T,R}$ for each measured distance lies in the confidence interval then the measurement cannot be considered unreliable and it is not rejected, otherwise the software diagnostics asks to reject the measurement. The following condition must be verified:

$$Q_{MIN} \le \left(\varepsilon_{T,R}\right)_i \le Q_{MAX}, \forall i \in \{1 \ldots N\} \tag{6.30}$$

where N is the number of network devices communicating with a given probe device.

6.6.1 Setup of Parameters

As usual, the risk level α is established by the user depending on the required level of performance of the system.

Table 6.3 Details of data analysis for standard deviation estimation of residuals

Sample dimension: $N_{TOT} = \sum_{j=1}^{M} N_j$	254
Mean estimate: $\hat{\mu}_{T,R} = \sum_{i=1}^{N}\sum_{j=1}^{M} \frac{(\varepsilon_{T,R})_{ij}}{\sum_{j=1}^{M} N_j}$	0.6 mm
Standard deviation estimate: $\hat{\sigma}_{T,R} = \sqrt{\sum_{i=1}^{N}\sum_{j=1}^{M} \frac{(\varepsilon_{T,R})_{ij}^2}{\sum_{j=1}^{M} N_j}}$	12.9 mm

Similarly to the two diagnostics models described in the previous sections, standard deviation $\sigma_{T,R}$ can be evaluated by applying the uncertainty composition law, or empirically, on the basis of experimental distance measurement.

Considering the second method, a set of M points randomly distributed in the measurement space $\xi \subseteq \mathbb{R}^3$ are randomly measured. For each measurement $(\varepsilon_{T,R})_{ij}$ is calculated (where $i = 1\ldots N_j$ and $j = 1\ldots M$). The number of residuals N_j may change due to the different number of distances, detected during each measurement.

Hypothesizing a normal distribution for all the random variables $(\varepsilon_{T,R})_{ij}$, i.e., $(\varepsilon_{T,R})_{ij} \sim N\left(0, \sigma_{T,R}^2\right)$. The standard deviation may be estimated as follows:

$$\hat{\sigma}_{T,R} = \sqrt{\sum_{i=1}^{N}\sum_{j=1}^{M} \frac{\left((\varepsilon_{T,R})_{ij} - 0\right)^2}{\sum_{j=1}^{M} N_j}} = \sqrt{\sum_{i=1}^{N}\sum_{j=1}^{M} \frac{(\varepsilon_{T,R})_{ij}^2}{\sum_{j=1}^{M} N_j}} \tag{6.31}$$

The obtained value $\hat{\sigma}_{T,R}$ is considered as the reference value for the test. Acceptance limits defined in Eq. 6.29 become:

$$\begin{aligned} Q_{MIN} &= z_{\frac{\alpha}{2}} \cdot \hat{\sigma}_{T,R} \\ Q_{MAX} &= z_{1-\frac{\alpha}{2}} \cdot \hat{\sigma}_{T,R} \end{aligned} \tag{6.32}$$

6.6.2 A Practical Case Study

Also for this kind of diagnostics an application example, referring to MScMS-I, is reported.

In order to estimate $\sigma_{T,R}$ the steps here below were followed:

- A sample of $M = 30$ points randomly measured by the probe was considered.
- The coordinates of each probe device were evaluated using the "mass-spring" localisation algorithm. A sample of 254 residuals were obtained.
- Normal distribution of residuals was tested using a chi-square test.

- Standard deviation of residuals was estimated using Eq. 6.31. The obtained result is $\hat{\sigma}_{T,R} = 12.9$ mm (see Table 6.3 for details).

The resulting 95% confidence interval for $\varepsilon_{T,R}$ is [−25.3; 25.3] mm. Therefore the generic point measurement cannot be considered unreliable if

$$\left|(\varepsilon^*_{T,R})_i\right| \leq 25.3\,\text{mm}\,\forall i \in \{1\ldots N\} \tag{6.33}$$

where $(\varepsilon^*_{T,R})_i$ is the calculated value of residual during a specific measurement.

Now consider the situation described in Fig. 6.10, where a probe device is located on point x_P. The distances measured according to the received signal are:

$$\begin{aligned} d_{R_1} &= 2104.8\,\text{mm} \\ d_{R_2} &= 1713.4\,\text{mm} \\ d_{R_3} &= 1831.4\,\text{mm} \\ d_{R_4} &= 2355.6\,\text{mm} \\ d_{R_5} &= 2215.2\,\text{mm} \end{aligned} \tag{6.34}$$

On the other hand, the distances measured according to the transmitted signal are:

$$\begin{aligned} d_{T_1} &= 2136.2\,\text{mm} \\ d_{T_2} &= 1695.5\,\text{mm} \\ d_{T_3} &= 1818.7\,\text{mm} \\ d_{T_4} &= 2357.7\,\text{mm} \\ d_{T_5} &= 2221.9\,\text{mm} \end{aligned} \tag{6.35}$$

Due to the reflection effect a significant difference between the two distances measured from node 1 can be detected. The following residual values are obtained:

$$\begin{aligned} (\varepsilon^*_{T,R})_1 &= 31.4\,\text{mm} \\ (\varepsilon^*_{T,R})_2 &= -17.9\,\text{mm} \\ (\varepsilon^*_{T,R})_3 &= -12.7\,\text{mm} \\ (\varepsilon^*_{T,R})_4 &= 2.1\,\text{mm} \\ (\varepsilon^*_{T,R})_5 &= 6.7\,\text{mm} \end{aligned} \tag{6.36}$$

Since the value of residual referring to node 1 is not included in the acceptance interval ([−25.3; 25.3] mm), the diagnostics test suggests to reject the measurement.

6.7 Further Remarks

Some of the innovative aspects of a distributed metrology system concern their on-line diagnostics tools. When dealing with measurement systems, the importance of a good diagnostics of produced measurements is crucial for applications in which errors can lead to serious consequences.

The diagnostics tools described in this chapter, all based on the concept of "reliability of a measurement", enable distributed systems to automatically reject measurements which do not satisfy a series of statistical acceptance tests with a given confidence coefficient.

For each measurement, if all these tests are satisfied at once, the measured result is considered acceptable. Otherwise, the measurement is rejected.

After rejection, the operator is asked to redo the measurement, changing the orientation/positioning of the probe or, if it is necessary, beacons arrangement in the system network.

Chapter 7
Methodologies for Performance Enhancing

7.1 The Practice of Error Correction

In accordance with the International Vocabulary of Metrology (VIM), a measurement error is defined as "the difference between the value obtained by the measurement and the actual value of the measured quantity (the so called "nominal value" or "true value", which is never known exactly)" (JCGM 200:2008 2008). Errors of measurement can be caused by many different sources, which are generally related to the technological features of the measuring instrument, the measurement procedure, the knowledge and experience of the operator, the characteristics of the measurand, the environment, and other effects. Consequently, a deep knowledge of the instrument and the context of measurement is needed in order to identify them in a thorough way.

For example, by considering a high precision instrument for dimensional measurement, such as a classical CMM, major error sources can be divided in different groups (Schwenke et al. 2008): kinematic, thermo-mechanical, loads, dynamic forces, motion control and control software.

Measurement errors can be divided into two main categories: random and systematic errors.

- *Random errors* are statistical fluctuations (in either direction) in the measured data caused by unknown and unpredictable changes in the experiment and due to limitations of the measurement device.
- *Systematic errors*, by contrast, are reproducible inaccuracies that are consistently in the same direction. Systematic errors are often due to a problem, relating to the measuring instrument or in the environmental conditions, which persists throughout the entire experiment. This kind of error is sometimes called "bias" and can be eliminated by applying a "correction" to compensate for an effect recognized during calibration—i.e., comparison to standards of known value (JCGM 200:2008 2008). Unlike random errors, systematic errors cannot be reduced by increasing the number of observations.

F. Franceschini et al., *Distributed Large-Scale Dimensional Metrology*,
DOI: 10.1007/978-0-85729-543-9_7,

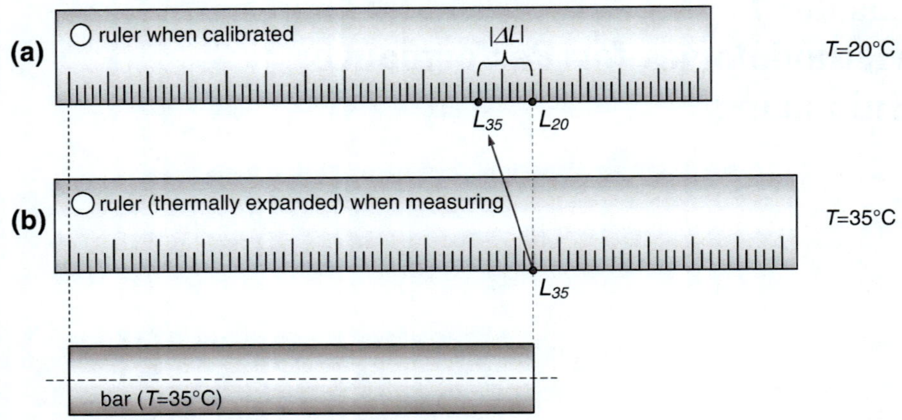

Fig. 7.1 Example of error correction. A ruler is used to measure the length of a *bar*. **a** the ruler's scale of measurement is calibrated at the temperature of 20°C, while **b** the bar is measured at the temperature of 35°C. This temperature difference will cause a thermal expansion of the scale with respect to the condition of calibration and, consequently, a systematic underestimation ($|\Delta L|$) of the measured bar length. Precisely, $|\Delta L| = L_{35} - L_{20}$, being L_{35} the (underestimated) result of the measurement taken at 35°C, and L_{20} the correct measurement result, assuming that the ruler would not be subject to thermal expansion

Metrological performance of every measuring instrument can be significantly enhanced by correcting systematic errors. Of course, a model able to convert a measurement value (affected by random and systematic errors) to the corresponding unbiased value of the measured quantity (affected by random errors only) is needed. For the purpose of clarity, a simplified example is presented.

Example A ruler made of aluminium is used to measure the length of a bar. The ruler scale of measurement has been calibrated (i.e., adjusted by comparison with a reference standard) at the temperature of 20°C. On the other hand, the bar is measured in its operating environment, at the temperature of 35°C. This temperature difference will therefore cause a thermal expansion of the ruler scale with respect to the condition of calibration and, therefore, a systematic underestimation of the measured bar length (see Fig. 7.1). For simplicity, we assume that other kinds of systematic and random errors can be neglected.

To determine the correct value of the bar length (L_{20}), Eq. 7.1 can be used

$$L_{20} = L_{35} \cdot [1 + \Delta T \cdot \alpha] = L_{35} \cdot [1 + (35 - 20) \cdot \alpha] \quad (7.1)$$

where L_{35} is the (underestimated) result of the measurement, taken at 35°C; L_{20} is the correct measurement result, assuming that the ruler would not be subject to thermal expansion; ΔT is the Temperature difference between the condition of measurement (35°C) and the condition of calibration (20°C); α is the Coefficient of linear thermal expansion , describing the tendency of aluminium (and most metals) to get longer as they are heated and shorter when they are cooled (ISO/TR 16015:2003 2003).

Thus, the model of Eq. 7.1 can be applied for correcting the systematic error (ΔL) due to thermal expansion (Eq. 7.2):

$$\Delta L = L_{35} - L_{20} = -L_{35} \cdot (35 - 20) \cdot \alpha \tag{7.2}$$

It can be noticed that this correction improves the instrument metrological performance, without any modification in the hardware.

Analogous correction models can be constructed for every kind of measuring instrument. As emerged from the example, typical correction activities can be:

- deep analysis of the instrument's technical features;
- identification of the major (systematic) error sources;
- construction of a theoretical/empirical correction model;
- model implementation and validation.

Since, correction models are strongly related to the technology and the characteristics of the measuring instruments, the more complex the instruments are, the more complex correction models are likely to be. In fact, in complex instruments many parts may interact, with many corresponding potential error sources. It is therefore important to identify the major ones considering not just their effects separately but their interaction as well, since it may play an important role in the total system accuracy (Schwenke et al. 2008; Montgomery 2008; Box et al. 1978).

Section 7.2 presents an organic analysis aimed at identifying and correcting some of the most influential systematic errors of MScMS-I. Section 7.3 contains some ideas about the construction of homologous models for other large-scale metrology distributed systems, i.e., MScMS-II and the iGPS^{TM}.

7.2 Performance Enhancing for MScMS-I

At present time, the accuracy of MScMS-I in determining the spatial position of the measured points is not very high. The basic reasons are that (1) MScMS-I (as well as MScMS-II) is still not a completely developed system, only a prototype, and (2) ultrasound technology can be subject to many error sources. Specifically:

- non punctiform dimension of US transceivers;
- positioning error of the network devices;
- speed of sound dependence on environmental conditions;
- ultrasound reflections or diffraction, due to the presence of obstacles or external uncontrolled sources of ultrasounds;
- battery charge level of US transceivers;
- use of amplitude threshold detection at receivers;
- errors related to the use of the optimization algorithm for trilateration;
- errors related to the portable probe geometry.

For a detailed description of these error sources, we refer the reader to (Franceschini et al. 2009c, 2010; Maisano and Mastrogiacomo 2010).

Some screening experiments have shown that the most important factors of systematic errors are those related to the sources of US signal attenuation (Franceschini et al. 2010). This is directly caused by the thresholding detection method of ultrasound, described in detail in Sect. 7.2.1 (Priyantha et al. 2005).

The goal of this section is to construct an experimental correction model to improve the metrological performance of MScMS-I. Section 7.2.1 presents a close examination of the system US transceivers. Sections 7.2.2 and 7.2.3 provide a description of the experimental set-up and results. The remaining subsections (Sects. 7.2.4, 7.2.5) focus the attention on the construction, implementation and validation of a two-factor experimental model to improve the accuracy in the MScMS-I' time-of-flight (TOF) measurements.

The proposed correction model can be successfully used at two different stages:

- during the network calibration (operation that has to be necessarily performed before measurements; see description in Chap. 5), so as to improve the localization of the network devices;
- during measurement procedure, so as to improve the accuracy in the localization of the wireless probe (see Chap. 3).

The experimental construction of the model is performed following an analytical statistical protocol, in which the TOF of the US signal is identified as the key factor (dependent variable) to be examined. Next, two independent variables affecting TOF (transceivers' distance and relative orientation) are analysed by means of the Design of Experiments (DoE) methodologies (Montgomery 2008; Box et al. 1978). Air *T* and *RH* are fixed and the presentation order is completely randomized to minimize order-of-testing effects. After showing that the two independent variables have a significant effect on the TOF measurement error, additional experiments are performed and a model is constructed by performing a linear regression based on the experimental results.

7.2.1 Analysis of the US Transducers

7.2.1.1 Basic Characteristics

US sensors are used in many fields. In general, the key features of ultrasound transducers change depending on the propagation medium (solid, liquid, or air). One of the most important applications of US transducers is distance measurement, in which the propagation medium of the acoustic signals is typically air. The common applications associated with distance measurement are presence detection, identification of objects, measurement of the shape and orientation of workpieces, collision avoidance, room surveillance, liquid level and flow measurement (Delepaut et al. 1986). The US ranging systems are traditionally low cost compared with other technologies, such as laser range measurement systems.

Unfortunately, these systems exhibit low accuracy, low reliability due to reflection of the transmitted signals, and limited range (Manthey et al. 1991).

US sensors provide high accuracy only in certain working contexts. Excellent performance can be achieved when measuring, for example, short fixed distances and under controlled environmental conditions (temperature and humidity). The most common technique for distance evaluation is by measuring the TOF of the US signal either from a transmitter to a receiver or using a single transceiver, which transmits the US signal and receives the corresponding reflected signal. Other factors influencing the performance of US sensors are the type of transducers and the signal detection method used (i.e., thresholding, envelope peak and phase detection, as discussed in Sect. 7.2.1.2). Different types of transducers can be employed depending on the specific application. Most of the commercially available air US transducers are ceramic-based and operate at 40 kHz. Transducers that operate at higher frequencies, such as at 200 kHz, are more limited and more expensive (Toda 2002).

In modern US distance measurement systems for industrial applications, piezoelectric transducers clearly dominate. The typical advantages are their compact rugged mechanical design, high efficiency, great range of operation temperature, and relatively low cost. Airborne ultrasound systems have been developed for many types of distance measurement using two possible techniques (Berners et al. 1995):

1. *Pulse echo*. A transducer emits a burst of ultrasound that bounces off any object in the path of the beam. The transducer then acts as a receiver for the reflected signal. A measurement of the time delay from transmission to reception determines the distance to the target.
2. *TOF*. A separate transmitter is pointed toward the receiver. Instead of relying on reflections, this system detects the direct transmission of the signal from transmitter to receiver. After measuring the TOF, the sensor's distance can be calculated knowing the speed of sound value, as shown in Eq. 3.8 (with regard to the notation in this equation, consider TOF $\approx$ TDoA $= \Delta t$) (Gustafsson and Gunnarsson 2003).

Cricket devices, being equipped with either a US transmitter and a receiver, implement the TOF technique. A complex problem, when using US transducers, is the choice of the characteristic parameters (typically, resonant frequency and bandwidth). For distance measurement with relatively high precision (few millimetres), transducers with a wide bandwidth are needed. The bandwidth is a measure of how rapidly a signal reaches the steady state. A signal at the receiver—which is obtained from transducers with a small bandwidth—slowly climbs from its beginning to its peak in time domain, which causes a relatively large transient time at the receiver. This behaviour is shown in Fig. 7.2 (Cheng and Chang 2007; Tong et al. 2001).

A second factor affecting the measurement accuracy is the transducer resonant frequency. With increased frequency (and, thus, reduced wavelength), a better resolution is achievable. Unfortunately, both the transducer bandwidth and the

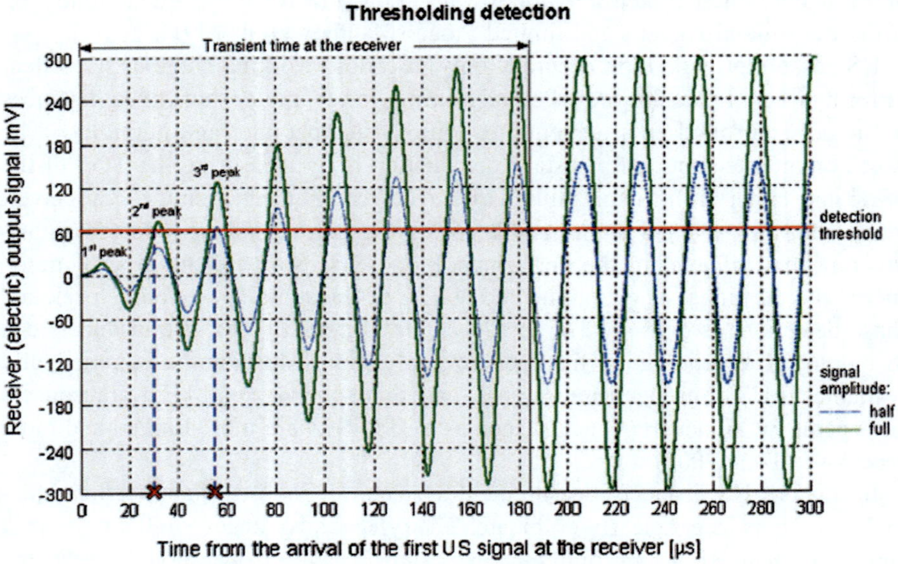

Fig. 7.2 Graphical representation of thresholding detection. A minimum number of cycles are necessary to bring the receiver to steady-state conditions (transient time at the receiver) (Johansson et al. 2006). The error in the distance measurement is dependent upon the received US signal amplitude (full or half in this specific case). The time taken for the received signal to reach the threshold is dependent on its amplitude (Franceschini et al. 2010). (with permission)

resonant frequency are directly correlated with ultrasound attenuation, and consequently they limit the detection range. In other words, considering the same US signal amplitude, the radiated signal amplitude at a given distance from the transmitter becomes smaller if its bandwidth and resonant frequency increase (Tong et al. 2001; Kažys et al. 2007). For this reason, the selection of US frequency and bandwidth is a compromise between accuracy and detection range.

The piezoelectric transducer adopted by Cricket devices is a low-cost general-purpose model (Murata MA40S4R, see Fig. 7.3a) with a relative wide bandwidth (see Fig. 7.3b), in which the central frequency is about 40 kHz. This working frequency is a tradeoff between accuracy (considering the single distances, it is around 1–2 cm) and detection range (up to 6–8 m) (Balakrishnan et al. 2003; Magori and Walker 1987).

The acoustic strength of the radiation from a flat transducer with "piston motion" (like the Crickets' US transducers) is generally angle-dependent because of the phase difference of waves from each point on the surface (Berners et al. 1995). The acoustic radiation is the integral sum of the waves from all points on the transmitter surface, and the propagation path difference from each point to a reference observation point has a "phase-cancelling effect" that leads to signal attenuation (Lamancusa and Figueroa 1990; Figueroa and Lamancusa 1992).

However, if the receiver is directly facing the transmitter at sufficient distance from it, the acoustic radiation from each point of the transducer surface does not

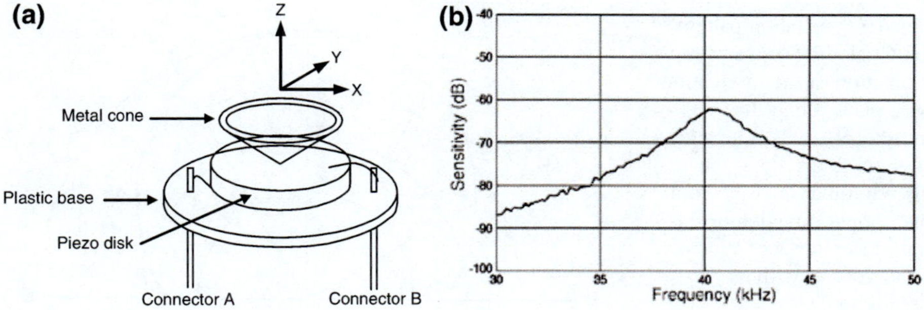

Fig. 7.3 Scheme of US device used in MScMS. **a** Internal construction of a Murata MA40S4R piezoelectric US transmitter/receiver. The dimensions of the piezomaterial cause the disk to resonate at a precise frequency (around 40 kHz). **b** Representation of the transmitter bandwidth by means of a frequency response plot (Priyantha et al. 2005). (with permission)

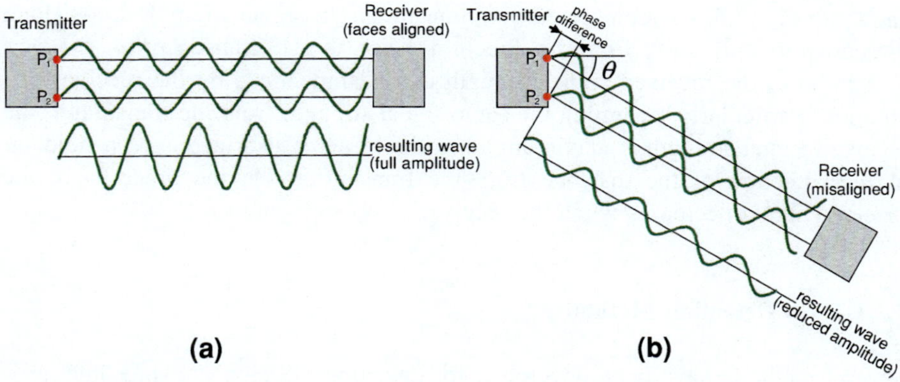

Fig. 7.4 US signal strength dependence on the transmitter angle (θ). The simplified illustration represents the interaction of the waves from two points (P_1 and P_2) on the transducer's surface. The resulting wave is given by the sum of the single waves. If the receiver is directly facing the transmitter (**a**), then the two individual waves are in phase and the resulting wave amplitude has the maximum value. If the transmitter is misaligned with respect to the receiver (**b**), then the resulting wave is attenuated because of a phase-cancelling effect due to the phase difference between the two individual waves (Lamancusa and Figueroa 1990; Franceschini et al. 2010). (with permission)

have a "phase-cancelling effect". This is because the distance from an arbitrary point on the transducer surface to the receiver becomes almost constant, and the difference is much smaller than the wavelength (Toda and Dahl 2005). On the other hand, if the transmitter is misaligned with respect to the receiver, then the US signal amplitude will be attenuated because of the disruptive interference of the different US signals from different surface points on the transmitter. This effect is represented by the simplified illustration in Fig. 7.4. The scheme considers the interaction of waves from two points on the transducer surface; the same principle can be extended to all the surface points.

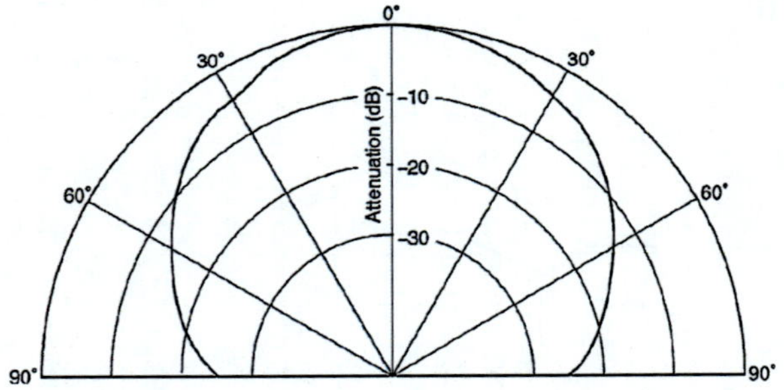

Fig. 7.5 Radiation pattern of the Cricket US transducer as a function of the orientation on a plane along the transducer's axis. The signal strength drops along the direction that is away from the normal direction to the transducer surface (Priyantha et al. 2005). (with permission)

An example of the resulting US transmitter radiation pattern as a function of the transmitter orientation with respect to the receiver ("misalignment angle") is shown in Fig. 7.5. As represented, the transmitter US signal strength drops along directions that are away from the direction facing the US transducer.

Similarly, the received signal strength can be influenced by the receiver orientation. Particularly, assuming the same signal strength from the transmitter, the received signal strength is maximum when the receiver's surface is perpendicular to the direction of the distance from the transmitter. On the other hand, the received signal decreases when the receiver's surface is angled.

7.2.1.2 US Detection Methods

Several methods have been developed for detecting US signals. Thresholding is the simplest and the most widely used and applies to any type of short-duration signal. In this method, which is used by Crickets, the receiver electric output signal is compared with a threshold level (65 mV for the Crickets) such that the arrival of the wave is acknowledged when the signal reaches this level (see Fig. 7.2). This method depends on the amplitude of the pulse received: the larger the signal amplitude, the smaller the time taken by the signal before reaching the threshold.

For example, in Fig. 7.2, when the signal has a full amplitude, the detection threshold is first exceeded by the second peak of the ultrasound waveform. When the waveform is attenuated by a factor of 0.50 (half amplitude signal), the detection threshold is first exceeded by the third peak of the ultrasound waveform. If the channel attenuation is quite significant, then it may cause the threshold to be exceeded a few periods late instead of just one period late. Considering that at 40 kHz the period is 25 μs, the error will be approximately in integer multiples of 25 μs. The error in the TOF evaluation results in an error in the distance estimation. The speed of sound is about 340 m/s, so one ultrasound time period corresponds to a distance of about $25 \times 340/1{,}000 = 8.5$ mm. In practice, the threshold can be exceeded by up to four periods late, so the distance overestimation can be up to 3–4 cm!

A modification of thresholding is "envelope peak detection," which may be called "adjustable thresholding". This method acknowledges the arrival of the signal when a maximum amplitude is detected; therefore, it does not depend upon the absolute magnitude of the pulse, only upon its shape. As a consequence, it is more accurate and robust than simple magnitude thresholding, where the acknowledge time can easily jump by one period.

Other more refined ranging methods are based on phase detection with fixed-frequency signals and with frequency modulated signals. These methods, however, require complex hardware and software. They use a digital signal processor to process the phase measurements to overcome the inherent range limitation of one wavelength (Mahajan and Figueroa 1999; Manthey et al. 1991; Hazas and Ward 2002; Parrilla et al. 1991; Kino 1987).

Recently, a lot of effort has been done to incorporate pulse compression techniques to the US sensory system to improve the accuracy in distance measurements using relatively simple hardware and software (Piontek et al. 2007).

7.2.2 Description of the Experiments

This section illustrates the experiments carried out for constructing the correction model. Experimental data were collected in the same operational conditions in which the correction model will be used (Box et al. 1978); more precisely, network devices were assumed to be parallel with respect to the devices to be localized. In the current practice, this condition is generally satisfied because network devices (C_1, …, C_n) are arranged on the ceiling, at the top of the measuring area and Crickets to be localized are generally mounted on the portable probe and oriented upwards (see Fig. 7.6). This configuration is a practical solution to obtain a good coverage and to maximize the measuring volume (Johansson et al. 2006). For a more general description of the problem of the measuring volume coverage, i.e., with no assumption of parallelism between the network devices and the device to localize, see Sect. 4.5.1.

In this configuration, the misalignment angles related to a generic network device (C_i) and the one related to the device(s) to be localized, with respect to their distance, is the same (θ_i in Fig. 7.6). Figure 7.7 illustrates the experimental setup:

- transmitter (T) and receiver (R) are positioned facing each other;
- the distance (*d*) between transceivers is known and represents the 1st factor of the factorial plan;
- transmitter face is parallel with receiver face, but they are not perpendicular with respect to the direction of the distance. A misalignment angle (θ) is introduced and represents the 2nd factor of the factorial plan.

As shown in Fig. 7.7, the reference point for determining the transceivers' distance and misalignment angle corresponds to the centre of each US transceiver

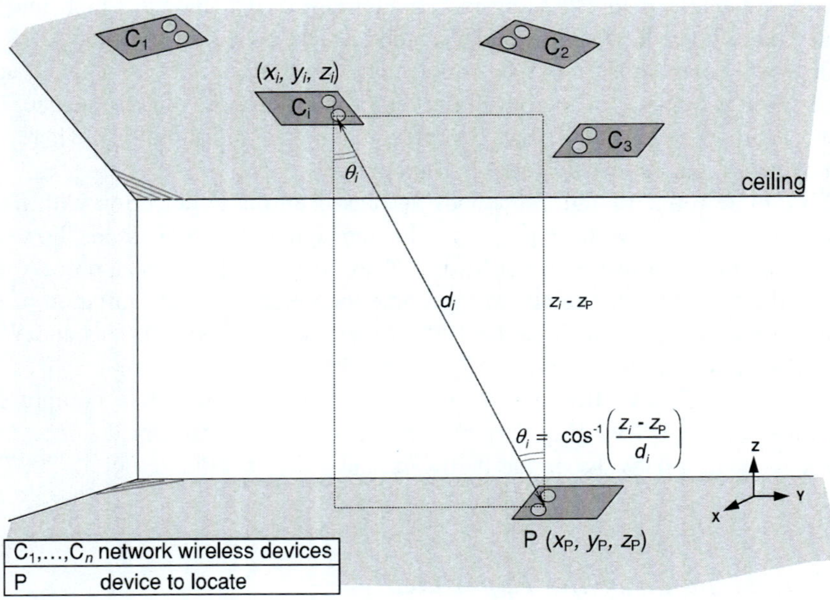

Fig. 7.6 Typical arrangement of the network devices (C_1–C_n). To obtain a good coverage, network devices are placed at the top of the measuring area, parallel to the ceiling. With this configuration, Cricket to localize (P) should be oriented upwards. The formula for calculating the misalignment angle (θ_i) is used in the iterative procedure for the TOF-Error correction, reported in Sect. 7.2.5.1 (Maisano and Mastrogiacomo 2010). (with permission)

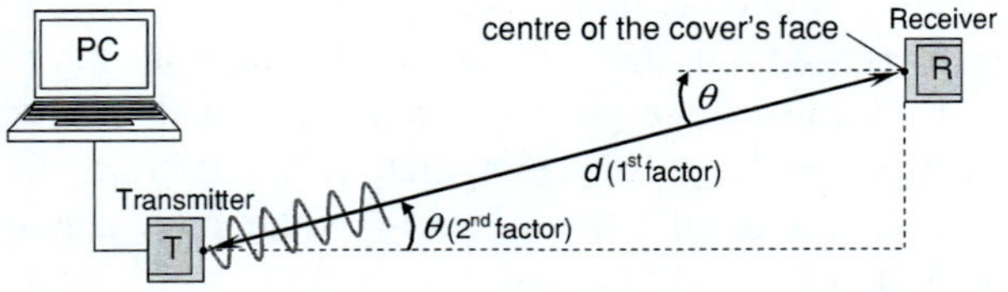

Fig. 7.7 Experimental set-up. Transmitter (T) and receiver (R) are positioned facing each other. Distance (d) between transceivers and transceivers' misalignment angle (θ) can be varied

cover's face. Distances and angles were measured using a set of calibrated reference bars and a precision goniometer (Magori 1994).

Experiments were organized in two steps:

1. *Exploratory experiments*. Based on a limited number of observations, this phase was aimed at investigating whether the two factors of interest have significant effects on the TOF measurements. To that purpose, an experimental factorial plan was built measuring TOF and changing the factors at different levels. Table 7.1 provides a summary of the factor level combinations.
2. *Detailed experiments*. The factor working domain and the number of observations are increased so as to build an empirical regressive model representing the effects of the two factors. Table 7.2 contains the list of observations considered in this phase.

Table 7.1 List of the TOF observations (functions of d and θ) considered in the exploratory phase

Factors	
1st—distance between transceivers (d) (mm)	2nd—Transceiver misalignment angle (θ) (°)
Levels	
1,000	0
2,000	15
3,000	30
	45

All the possible 3 × 4 = 12 different combinations are carried out in random order

For each combination, TOF measurements are repeated 50 times

All the 12 combinations above are replicated 3 times; consequently, the total number of combinations is 36. The test sequence is randomized using the random number generator provided by Minitab®

Table 7.2 List of the TOF observations (functions of d and θ) considered in the detail phase

Factors	
1st—distance between transceivers (d) (mm)	2nd—transceiver misalignment angle (θ) (°)
Levels	
500	0
1,000	15
1,500	30
2,000	45
2,500	60
3,000	
3,500	
4,000	

All the possible 8 × 5 = 40 different combinations are carried out in random order

For each combination, TOF measurements are repeated 50 times

For each of the combinations in Tables 7.1 and 7.2, 50 repeated measurements of the TOF are performed.

The response variable considered in the factorial plan is the TOF-Error, which is used as an indicator of the error in TOF evaluation and defined as follows:

$$\text{TOF-Error} = \text{TOF} - \text{Expected-TOF} \tag{7.3}$$

being:

TOF	TOF measured by pair of Crickets;
Expected-TOF	Defined as d/s, where d is the known distance between transceivers and s is the speed of sound in the experimental conditions (with air temperature $T = 21°\text{C}$ and relative humidity $RH = 27\%$, $s \approx 344$ m/s)

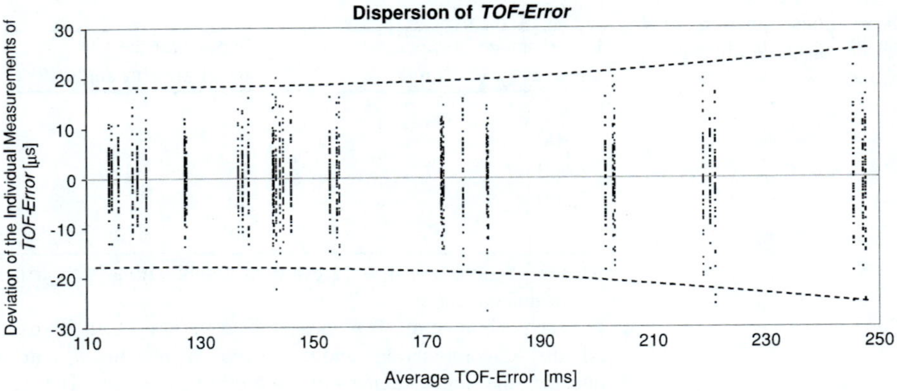

Fig. 7.8 TOF-Error deviation versus average TOF-Error. For each of the 36 factors combinations, variables are calculated using the corresponding 50 repeated TOF-Error measurements (Maisano and Mastrogiacomo 2010). (with permission)

7.2.3 Analysis of the Experimental Results

7.2.3.1 Results of the Exploratory Experiments and Factorial Analysis

Analysing the output of the exploratory factorial experiments, it can be noticed that TOF-Error dispersion changes depending on the average of the TOF-Error value. This behaviour is illustrated in Fig. 7.8. For each of the 36 factorial plan combinations, the average TOF-Error and the respective deviations—calculated using the corresponding 50 repeated measurements—are plotted. In general, the larger the average TOF-Error value, the larger the individual measurement dispersion. The non homogeneity of the TOF-Error variance is also tested through the Levene's statistical test, at $p < 0.05$.

Since the assumption of homogeneity of TOF-Error variances is violated, the Analysis of Variance (ANOVA) cannot be properly applied, in order to verify whether factors (d, θ) have a significant effect on the response (TOF-Error) (Box et al. 1978). The usual approach to dealing with non-homogeneous variance is to apply a variance-stabilizing transformation. In this approach, the conclusions of the analysis of variance will apply to the transformed populations. The most common transformation is the exponential $y* = y^{\lambda}$, where λ is the parameter of the transformation (Godfrey et al. 2006; Zhang et al. 2005).

The parameter λ is selected by the Box-Cox method. The experimenter can analyse the data using $y*$ as the transformed response (hereafter, it will be identified as "*Transformed TOF-Error*"). In our specific case, the obtained transformation parameter is $\lambda = 0.17$. Applying the Levene's test to the transformed response, the resulting variance no longer violates the test's null hypothesis of homogeneity. To construct a model in terms of the original response, the opposite change of variable—$(y = y^*)^{\frac{1}{\lambda}}$—is performed.

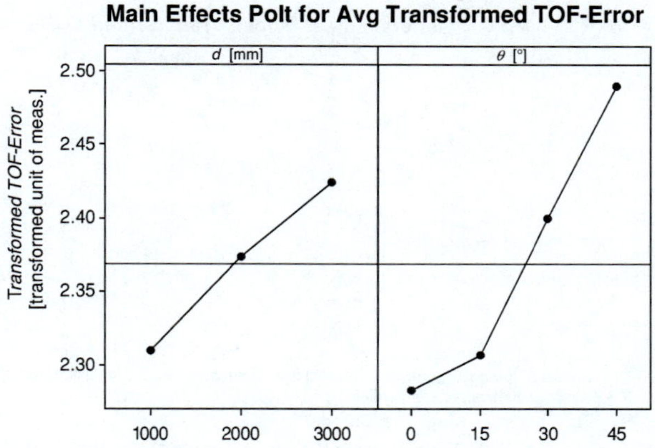

Fig. 7.9 Main effect plot for means, related to the two examined factors: *d* (transmitter–receiver distance) and *θ* (misalignment angle) (Maisano and Mastrogiacomo 2010). Graph is created using the Minitab® statistical software. (with permission)

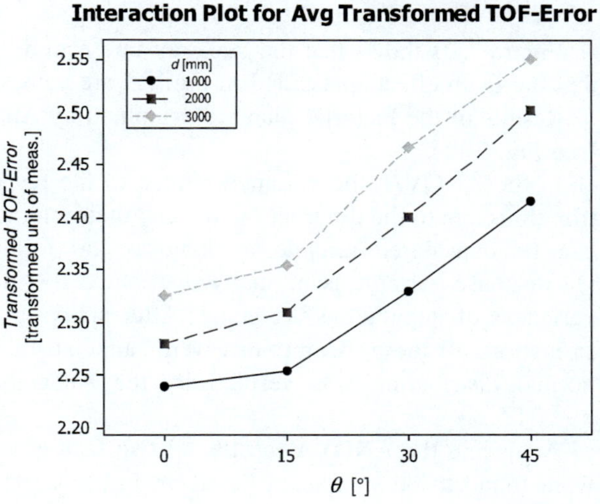

Fig. 7.10 Interaction plot for Transformed TOF-Error, considering the two factors (*d* and *θ*) (Maisano and Mastrogiacomo 2010). Graph is created using the Minitab® statistical software. (with permission)

The Main Effects Plot, representing the effect of the single examined factors on the TOF-Error, is shown in Fig. 7.9.

The points in the plot are the means of the response variable at the various levels of each factor (for each level of the examined factor, the mean is calculated averaging all the responses obtained changing the remaining factor). A reference line is drawn at the grand mean of the response data. This kind of plot is useful for comparing magnitudes of main effects. The qualitative result is that both d and θ have an important effect on the TOF-Error.

In order to qualitatively judging the presence of interactions among the two factors, an Interaction Plot is constructed in Fig. 7.10. Interaction between two factors is present when the response at a factor level depends upon the level(s) of the other factor. Parallel curves in an interactions plot indicate no interaction. The greater the departure of the curves from the parallel state, the higher the degree of interaction (Box et al. 1978).

General Linear Model: *Transformed TOF-Error* versus *d* and θ

Factor	Type	Levels	Values
d	fixed	3	1000; 2000; 3000 mm
θ	fixed	7	0°; 15°; 30°; 45°

Analysis of Variance for *Transformed TOF-Error*

	Source	DF	Seq SS	Adj SS	Adj MS	F	P
single factor effect	d	2	0.078990	0.078990	0.039495	4859.41	0.000
single factor effect	θ	3	0.243035	0.243035	0.081012	9967.54	0.000
interactions between factors	$d*\theta$	6	0.003364	0.003364	0.000561	68.97	0.000
	Error	24	0.000195	0.000195	0.000008		
	Total	35	0.325583				

DF degrees of freedom, *Seq SS* sequential sum of squares, *Adj SS* adjusted sum of squares, *Adj MS* adjusted mean squares, *F* Fisher's random variable, *P* *p*-value

Fig. 7.11 ANOVA applied to the (transformed) response of the factorial plan. This table is the output of a two-way ANOVA, carried out using Minitab® statistical software

Figure 7.10 shows that the two-way interactions are not very pronounced and that the main effects presented in Fig. 7.9 are consistent within each factor level.

Results of the factorial plan are examined by Analysis of Variance (ANOVA) (see Fig. 7.11).

In the ANOVA, the variance related to the response is partitioned into contributions due to the different factors and their interactions. Results of an ANOVA can be considered reliable as long as the following assumptions are met: (1) response variable is normally distributed, (2) data are independent, and (3) variances of populations are equal. After applying the Box-Cox response transformation, all these assumptions were satisfied. In particular, the assumption of normal distribution was verified by the Anderson–Darling normality test at $p < 0.05$.

Analysing the ANOVA results, all two factors and their two-way interactions were found to be significant based on Fisher's test at $p < 0.05$. With regard to single factors, both d and θ have an important effect. This is consistent with the Main Effects Plot of Fig. 7.9. With regard to the factor interaction, it is statistically significant too ($p < 0.05$). Thus, it can be stated that the composition of large misalignment angles (θ) and large distances (d) produces TOF-Errors, which are larger than those obtained adding the effects of the single factors, taken separately.

7.2.3.2 Results of the Detailed Experiments

Results of detailed experiments are graphically represented in Figs. 7.12 and 7.13. They represent the average TOF-Error and the corresponding standard deviation (calculated, for each combination of factors, using the 50 repeated measurements) depending on d and θ. From the two graphs in Fig. 7.12, we can notice that TOF-Error increases with an increase in d and θ. Again, TOF-Error is always positive, because of the TOF overestimation due to the signal attenuation (which is proportional to d and θ). In particular, the relationship between TOF-Error and

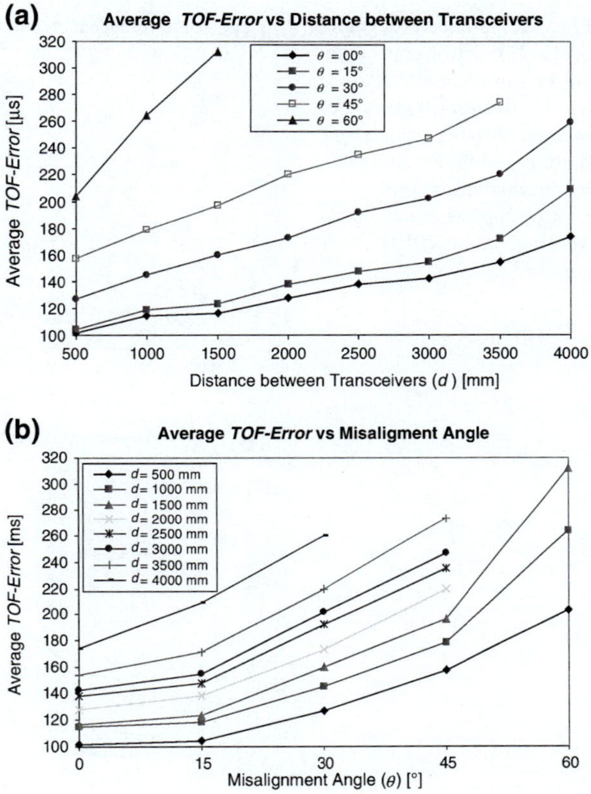

Fig. 7.12 Average value of the TOF-Error depending on the misalignment angle (θ), for different transmitter–receiver distances (d). **a** and **b** are two different graphic representations of this relationship (Maisano and Mastrogiacomo 2010). (with permission)

d appears approximately linear, while the relationship between TOF-Error and θ appears approximately quadratic. Also, it can be noticed that TOF-Error measurements cannot be performed when the two factors have both large values—i.e. when $\theta = 45°$ and $d > 3{,}500$ mm, and when $\theta = 60°$ and $d > 1{,}500$ mm (see also Table 7.3). In fact, in all these conditions transmitter and receiver are not able to communicate because of the strong signal attenuation (receiver beyond the transmitter's cone of communication).

Figure 7.13 also shows that the TOF-Error standard deviation tends to increase while increasing the received US signal attenuation; here follows an explanation of this behaviour. The amplitude of US signals at the transmitter is not perfectly stable but rather has a certain natural variability derived by many sources, such as power and control supply, air temperature, air turbulence, signal bandwidth. The envelope of the US signal amplitude at the receiver is included within an uncertainty band (highlighted in grey in Fig. 7.14). Considering signals with different amplitudes and assuming the uncertainty bandwidth to be the same, the larger the transient slope, the lower the TOF-Error uncertainty (U_1 and U_2 in Fig. 7.14).

Of course, the behaviour previously described is a consequence of the use of the thresholding detection method at the US receivers. The Cricket's accuracy could be improved by implementing a more refined ultrasound detection method that is

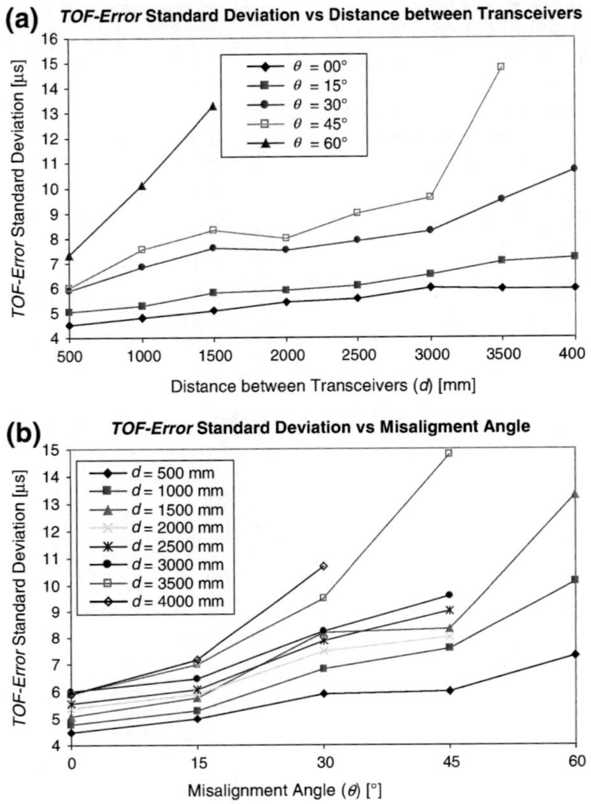

Fig. 7.13 Standard deviation of the TOF-Error depending on the misalignment angle (θ), for different transmitter–receiver distances (d). **a** and **b** are two different graphic representations of this relationship (Maisano and Mastrogiacomo 2010). (with permission)

Table 7.3 Communication between transceivers, depending on factors d and θ

	θ (°)				
	0	15	30	45	60
d (mm)					
500	✔	✔	✔	✔	✔
1,000	✔	✔	✔	✔	✔
1,500	✔	✔	✔	✔	✔
2,000	✔	✔	✔	✔	×
2,500	✔	✔	✔	✔	×
3,000	✔	✔	✔	✔	×
3,500	✔	✔	✔	✔	×
4,000	✔	✔	✔	×	×
4,500	✔	✔	✔	×	×

For some particular combinations of the two factors, transceivers are not able to communicate and, consequently, the experimental table cannot be completely filled. To be precise, measurements can be performed only for 37 of 45 (9×5) combinations. When the two factors have both large values—i.e., when $\theta = 45°$ and $d > 3{,}500$ mm, and when $\theta = 60°$ and $d > 1{,}500$ mm—measurements are not feasible

✔ measurement performed

× measurement not feasible

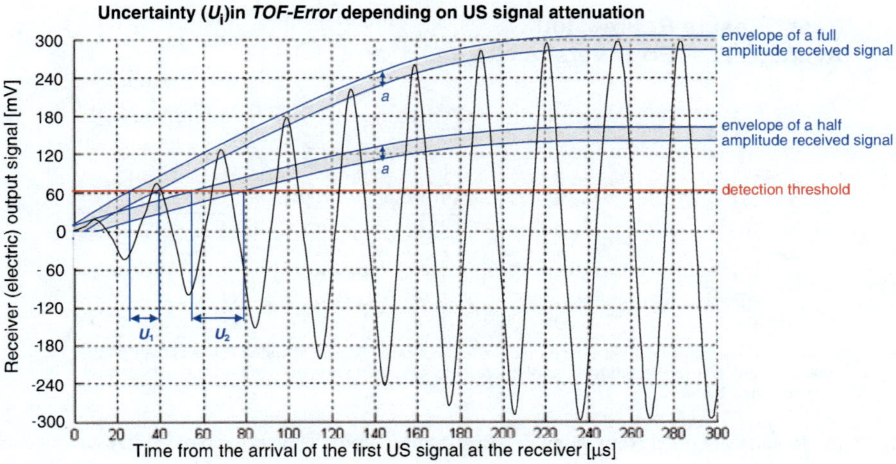

Fig. 7.14 Considering the same uncertainty (a) in the receiver voltage signal, the more attenuated the signal, the larger the TOF uncertainty (U_1 and U_2) (Franceschini et al. 2010). (with permission)

able to calculate the precise instant in which the US signal is received (Piontek et al. 2007).

7.2.4 Model Construction

In this section we propose an experimental regressive model to link TOF-Error with θ and d. The total number of observations that were used to construct the model is given by 37 combinations with 50 repetition per combination, which corresponds to 1,850 TOF measurements (see Table 7.3).

Analyzing the graph patterns in Fig. 7.12 and based on the knowledge of the physical phenomenon investigated, a second order polynomial model was chosen. For more information on the typical relationships between TOF-Error and d, and TOF-Error and θ, we refer the reader to the relevant literature (Manthey et al. 1991; Figueroa and Lamancusa 1992; Magori 1994).

Note that the chosen model makes it possible to evaluate the factors interaction:

$$\text{TOF-Error} = K_1 + K_2 \cdot d + K_3 \cdot \theta + K_4 \cdot d^2 + K_5 \cdot \theta^2 + K_6 \cdot d \cdot \theta \qquad (7.4)$$

With the support of the Minitab® Best-Subsets tool, we find that the terms with coefficients K_3 and K_4 have slightly influential contributions. In fact, considering several competing multiple regression models of order not larger than two (see Fig. 7.15), the model with the three terms (d, θ^2 and $d{\cdot}\theta$) is the one with the Mallows' Cp (4.3) closest to the number of predictors plus the constant (4). In general, Mallows' Cp is used in statistics to assess the fit of a regression model that

Best Subsets Regression:
Average *TOF-Error* versus d, θ, d^2, θ^2 and d·θ

Vars	R^2	R^2 (adj)	Mallows' *Cp*	S	(K_2) *d*	(K_3) *θ*	(K_4) d^2	(K_5) θ^2	(K_6) *d·θ*
1	72.1	71.2	155.6	28.715					X
1	55.4	54.0	266.8	36.307	X				
2	92.6	92.1	21.4	15.052	X			X	
2	90.1	89.5	37.7	17.359			X	X	
3	95.0	94.5	**4.3**	11.785	X			X	X
3	94.3	93.7	12.2	13.449		X	X	X	
4	95.7	95.2	7.3	12.566	X	X		X	X
4	95.5	94.8	6.6	12.167	X	X	X	X	
5	95.8	95.0	6.4	11.922	X	X	X	X	X

Vars is the number of variables or predictors in the model. R^2 and adjusted R are converted to percentages. S is the standard error of the regression. Predictors that are present in the model are indicated by an X.

Fig. 7.15 Results obtained from Minitab® Best-Subsets tool. The above table suggests that the model with the three terms d, θ^2 and $d \cdot \theta$ is relatively precise and unbiased because its Mallows' *Cp* (4.3) is closest to the number of predictors plus the constant (4) (Kino 1987)

has been estimated using ordinary least squares. It is applied in the context of model selection, where a number of predictor variables are available for predicting some outcome, and the goal is to find the best model involving a subset of these predictors. As anticipated, the best model is the one with the Mallows' *Cp* closest to the number of predictors plus the constant (Mallows 1973). In this specific case, this fact was also confirmed by an initial regression, based on the model in Eq. 7.4, in which the contribution of the terms d, θ^2 and $d{\cdot}\theta$ appeared to be secondary.

As a consequence, terms with coefficients K_3 and K_4 were removed from the model and a new second order model, representing a compromise solution between best-fitting and reduction of the number of predictors was constructed using Eq. 7.5.

$$\text{TOF-Error} = K_1 + K_2 \cdot d + K_5 \cdot \theta^2 + K_6 \cdot d \cdot \theta \tag{7.5}$$

The model requires the information about the distance and the misalignment angle related to each pair of Cricket devices. Being linear with respect to d and quadratic with respect to θ, the model well represents the graph patterns in Fig. 7.12. It is important to note the presence of the last term ($K_6{\cdot}d{\cdot}\theta$), which accounts for the interaction between the two factors.

Since the variance of the response variable (TOF-Error) is not homogeneous, a simple linear regression is not perfectly suitable. In particular, heteroscedasticity may have the effect of giving too much weight to subset of the data where the error variance is larger, when estimating coefficients. To reduce standard error associated with coefficient estimates, in regression in which homoscedasticity is violated, a common approach is to weight observations by the reciprocal of the estimated point variance (Box et al. 1978; Zhang et al. 2005; Jiménez and Seco 2005). For each observation, the variance is calculated using the 50 repetitions associated to the

(Weighted) Regression Analysis: *TOF-Error* versus *d*, θ and *d*· θ

The regression equation is:

$TOF\text{-}Error = 84.6 + 0.0207 \cdot d + 0.0314 \cdot \theta^2 + 0.000336 \cdot d \cdot \theta$

Predictor	Coef	SE Coef	T	P
Constant	84.6410	0.6040	140.14	0.000
d	0.0206548	0.0002960	69.78	0.000
θ^2	0.0314441	0.0004298	73.15	0.000
$d \cdot \theta$	0.00033568	0.00001144	29.35	0.000

S = 1.65608   R-Sq = 93.9%   R-Sq(adj) = 93.8%

Analysis of Variance

Source	DF	SS	MS	F	P
Regression	3	71072	23691	8637.96	0.000
Residual Error	1696	4651	3		
Total	1699	75723			

Source	DF	Seq SS
d	1	18037
θ^2	1	50672
$d \cdot \theta$	1	2362

DF degrees of freedom, *SS* sum of squares, *MS* mean squares, *F* Fisher's random variable, *P* *p*-value

Fig. 7.16 Results of the (weighted) regression analysis. Analysis is carried out using Minitab® Statistical software

corresponding factor combination (numerical values of the σ related to each factor combination are reported in Fig. 7.13). The final regression equation is:

$$\text{TOF-Error} = 84.6 + 0.0207 \cdot d + 0.0314 \cdot \theta^2 + 0.000336 \cdot d \cdot \theta \qquad (7.6)$$

In Eq. 7.6, TOF-Error, d and θ are respectively expressed in μs, mm and degrees (°).

This model can be useful for correcting the systematic error in TOF measurements. Given that the variation in the response standard deviation is not very large, it emerged that Eq. 7.6 is not very dissimilar to the result that would be obtained by a simple (non weighted) linear regression.

The regression output is quantitatively examined by an ANOVA (see Fig. 7.16). Based on t test at $p < 0.05$, it can be sentenced that all the terms in Eq. 7.6 are significant. Examining the residuals, they can be considered as randomly distributed by the Anderson–Darling normality test at $p < 0.05$. The model fits well with experimental data.

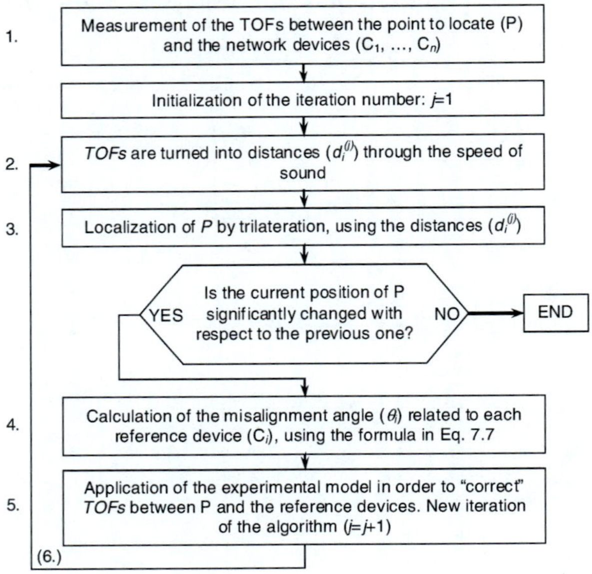

Fig. 7.17 Flowchart related to the iterative procedure for the on-line implementation of the model (Maisano and Mastrogiacomo 2010). (with permission)

7.2.5 Model Implementation and Validation

7.2.5.1 On-line Model Implementation

The goal of the correction model is to enhance the accuracy in the localisation of positioning target(s). Similarly to the diagnostics tools discussed in Chap. 6—the model is implemented on-line, both during network localisation and measurements.

It has to be noticed that, in order to correct the TOF between each network device and the positioning target (basically using Eq. 7.6), the mutual distance (d) and misalignment angle (θ) must be known. In turn, d and θ can be calculated only by knowing the network device position (a priori known) and the positioning target coordinates (not known a priori). Therefore, in order to implement the model, we studied an iterative procedure based on the gradual refinement of the positioning target position. This procedure is also based on the assumption that network devices are generally parallel to the device to be localized. Procedure consists of the following basic steps (see flowchart in Fig. 7.17):

1. TOFs between the device to localize (P) and the network devices with known position (C_1, …, C_n) are measured and then the iteration number (j) is initialized to 1.
2. Corresponding distances are calculated as: $d_i^{(j)} = \mathrm{TOF}_i^{(j)} \cdot s$. Superscript "($j$)" indicates that the jth iteration is considered. Since $\mathrm{TOF}_i^{(j)}$ is generally overestimated because of the attenuation, $d_i^{(j)}$ will of course result to be overestimated.

3. Device P (with coordinates $x_P^{(j)}$, $y_P^{(j)}$ and $z_P^{(j)}$) is localized through a trilateration, using the distances from at least three reference devices with known position (x_i, y_i, z_i).
4. Misalignment angles ($\theta_i^{(j)}$) between device P and each of the network devices, with which it communicates, are calculated. Thanks to the fact that devices have approximately parallel faces, we can use Eq. 7.7 (see Fig. 7.6):

$$\theta_i^{(j)} = \cos^{-1}\left(\frac{z_i - z_P^{(j)}}{d_i}\right) \tag{7.7}$$

5. Correction of the TOFs associated to each of the network devices using the following formula:

$$\text{TOF}_i^{(j+1)} = \text{TOF}_i^{(1)} - \text{TOF-Error}_i^{(j)} \tag{7.8}$$

   where TOF-Error$_i^{(j)}$ (function of $d_i^{(j)}$ and $\theta_i^{(j)}$) is calculated using the empirical formula in Eq. 7.6.
6. New estimation of the distances ($d_i^{(j\,+\,1)}$) between P and the devices with known position, and repetition of the procedure (steps 2–5, replacing superscripts j with $j + 1$).

The same procedure can be iterated until changes in the calculated position of P are not significant. Conventionally, this condition is reached when the distance between the current position of P and the position in the previous iteration is smaller than 1 mm. A qualitative example of application of the procedure is given by the representation scheme in Fig. 7.18.

This algorithm is designed to guarantee the convergence to a stable solution. Typically, it was experienced that no more than three iterations were necessary for the algorithm to converge. Using a standard PC, the total time to complete the procedure is not larger than 0.2 s, therefore it is fully compatible with the Crickets' TOF measurement sampling period.

7.2.5.2 Model Validation

Additional measurements have been performed so as to experimentally validate the empirical regressive model in conditions that are representative of the typical working environment. At this stage, a network of devices and a set of devices to be localised within the measuring volume were considered. It is important to stress that the model was based on the assumption that all the Crickets have parallel faces. Unfortunately, this condition can not be perfectly satisfied in a real measurement context, for two reasons:

1. Network devices are not perfectly parallel to each other. This condition would slow down and complicate the manual arrangement of devices and compromise

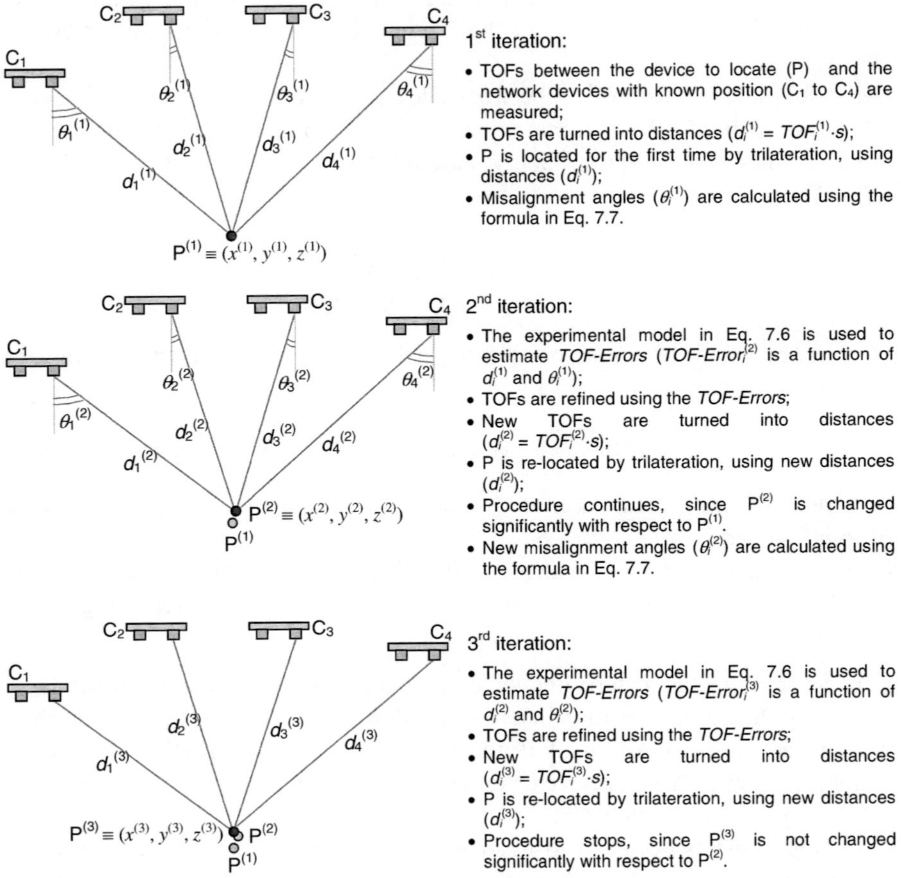

Fig. 7.18 Representation scheme of the iterative procedure for the experimental model implementation. It is assumed that three iterations are sufficient for the algorithm to converge. It can be seen that the position of the device to localize (i.e., $P^{(1)}$, $P^{(2)}$ and $P^{(3)}$) is gradually refined

the MScMS' easy start-up (Franceschini et al. 2009c). We found that, in a typical (quick) arrangement, misalignments of network devices with respect to the "ideal" parallelism condition are included within 3° (Franceschini et al. 2008).

2. Devices to be localised, which are mounted on the mobile probe, are not necessarily parallel to network devices. In fact, during the measurement task it is very difficult to keep the mobile probe always horizontal and facing network devices, due to the shape of the measured object. We confirmed that mobile probe's misalignment with respect to the "ideal" orientation can be up to 10–15° (Franceschini et al. 2008).

Thus, to test the efficiency of the proposed model in realistic measurement conditions, small misalignments—of the same amount of the ones mentioned before—were deliberately introduced both for network devices and devices to be measured. The test was organised as follows:

- A limited indoor measuring volume of about 24 m^3 (4 × 3 × 2 m) was considered. Eight network devices were distributed at the top of the volume, with parallel faces and a planar density of about 0.7 devices/m^2 (see Fig. 7.19). The rough position of each device was randomly set using a random number generator. Then, coherently with the previous experiments (Fig. 7.7), the "reference" position of each device—that is to say the Cartesian coordinates of the point coinciding with the centre of the transceiver cover's face—was calculated using a laser-tracker. Because of the transceiver's relatively small dimensions, it was difficult to measure the cylindrical cover touching it directly with the laser-tracker (cat's-eye) retroreflector. Therefore, measurements were performed using a support "cap"—i.e., an auxiliary component consisting of a hollow cylinder surmounted by a hemisphere. More precisely, the internal hole of the cap fits the US transducer's cover, so that—when the transducer is "capped"—the centre of the cover's face coincides with the centre of the cap hemisphere (see Fig. 7.20b). The cap is made of aluminium and manufactured using a CNC turning lathe with typical uncertainty around few hundredths of millimetre.
  The measurement procedure consists in:
  - "capping" the transducer with the support cap;
  - measuring several points (i.e. four or more), which are uniformly distributed on the hemisphere surface, using the laser-tracker retroreflector (see Fig. 7.20a);
  - determining the coordinates (x, y, z) of the hemisphere centre through a standard optimization algorithm.
    A preliminary uncertainty budget is constructed considering the uncertainty related to (1) the dimensional features of the support cup; (2) the laser-tracker measurement of the points on the cap hemispheric surface; (3) the algorithm to determine the centre of the hemisphere, using the previous points. The result is that the uncertainty associated to the coordinates of the point to measure is reasonably smaller than one millimetre, which is 1–2 orders of magnitude more accurate than the Cricket distance measurements (Ronchetti and Staudte 1994; Cross et al. 1998; JCGM 100:2008 2008).
- A Cricket device was placed next to twenty representative points that are positioned within the measuring volume, with the face parallel to the network devices. The rough position of each point was randomly set using a random number generator. Next, the "reference" position was measured by a laser-tracker with the same procedure seen before. For each of the twenty points, eight TOFs from the corresponding network devices were collected replicating the individual measurements five times. In practice, the device to be localized was moved and repositioned before each measurement, with the aim of reproducing the usual measurement conditions. Thus, there are 20 × 8 × 5 = 800 total TOF measurements (number of measured points × number of

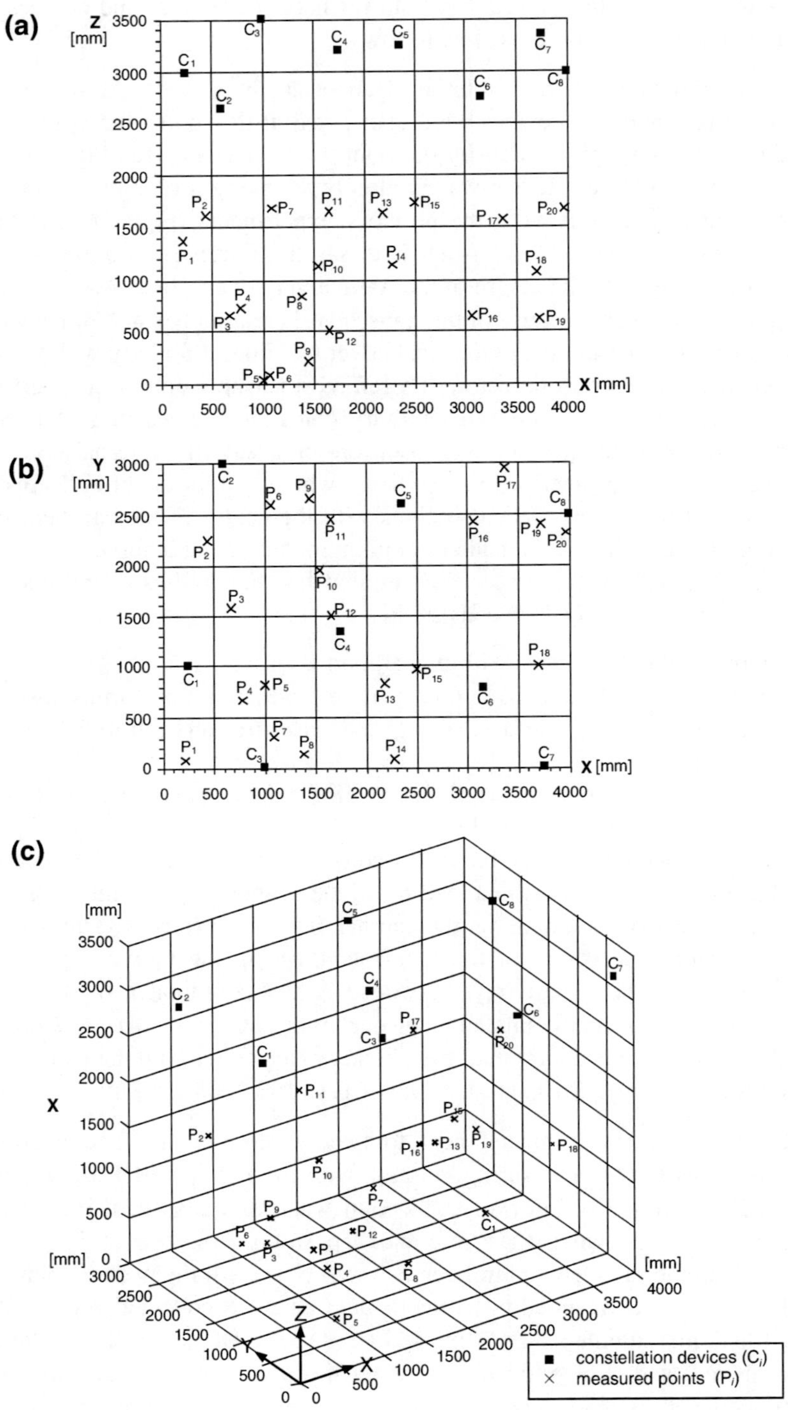

Fig. 7.19 Representation of the experimental set-up used for the model validation experiment. **a** XZ plane view, **b** XY plane view, **c** 3D view. The measuring volume contains eight network devices (*filled square*) and twenty measured points (*cross*), which are randomly positioned within the measuring volume (Maisano and Mastrogiacomo 2010). (with permission)

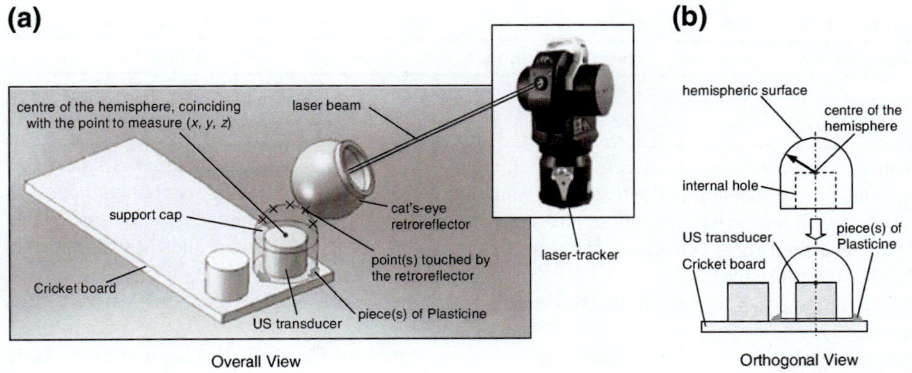

Fig. 7.20 Representation scheme of the procedure to calculate the coordinates of the point in the middle of the face of the US transducer cover (Maisano and Mastrogiacomo 2010). (with permission)

network devices × replications). These TOFs were turned into corresponding distances applying the iterative procedure seen in Sect. 7.2.5.1. Then, distance values were compared with the corresponding nominal values—derived using the laser-tracker reference positions—so as to calculate the resultant error. Results obtained by the application of the experimental model were compared (1) with the results obtained by applying a first order one-factor model, which was proposed by Moore et al. (2004) and implemented in a previous version of the Cricket firmware (see Eq. 7.9), and (2) with the results obtained with no correction.

$$d_i = 49.671 + 0.00096 \cdot s \cdot \mathrm{TOF}_i \tag{7.9}$$

In Eq. 7.9, d_i, s and TOF_i are respectively expressed in mm, m/s and μs.

Using the one-factor model in Eq. 7.9, distances (d_i) can be calculated from the corresponding TOF_is. Differently from the two-factor model, this model only accounts for the attenuation effect due to d, but does not consider the effect of θ.

Summarizing, distance error is calculated in the three following situations: (1) application of the two-factor empirical model; (2) application of the one-factor model; (3) no correction. Next, the average value ($\mu_{d\text{-}error}$) and the standard deviation ($\sigma_{d\text{-}error}$) related to distance error are calculated. Results are reported in Table 7.4. It is relevant to emphasize that these errors are not overall uncertainties for the system, because they are not achieved when all the error sources are combined in an uncertainty budget , also including traceable calibration uncertainties of the reference artefacts (Cross et al. 1998; JCGM 200:2008 2008).

It can be noticed that the two-factor correction model, compared to the one-factor model, makes it possible to reduce the dispersion in the distance evaluation considerably. Precisely, reduction is larger than 40%—i.e., $(9.8 - 5.6)/9.8$. The price to pay is that the two-factor model is based on the assumption that network

Table 7.4 Results of validation experiments with regard to the Cricket distance error

Distance error	Two-factor experimental model	One-factor experimental model	No correction
$\mu_{d\text{-}error}$ (mm)	−0.3	0.6	70.5
$\sigma_{d\text{-}error}$ (mm)	5.6	9.8	15.9

Notice that the two-factor correction model, compared to the one-factor model, makes it possible to reduce the dispersion in the distance evaluation considerably. Moreover, results obtained with no correction are very poor, both in terms of centering ($\mu_{d\text{-}error} = 70.5$ mm, due to the systematic TOF measurement overestimation) and dispersion ($\sigma_{d\text{-}error}$) (Maisano and Mastrogiacomo 2010with permission).

$\mu_{d\text{-}error}$ and $\sigma_{d\text{-}error}$ are calculated considering 800 individual distance evaluations, performed in random order

Reference distances are obtained using a laser-tracker (see Fig. 7.20a), with a measurement uncertainty one-two orders of magnitude smaller than Crickets'

devices are parallel with respect to the device to localize (see Fig. 7.6). Also, from Table 7.4 we can see that results obtained with no correction are very poor, both in terms of centering (large $\mu_{d\text{-}error}$, due to the systematic TOF measurement over-estimation) and dispersion ($\sigma_{d\text{-}error}$). In the latter case, the reduction of the dispersion is larger than 60%—i.e., $(15.9 - 5.6)/15.9$.

Not depending on the network device density, but only depending on the TOF measurements between device to localize and network devices, these results can be extended to networks with different density.

It is important to remark that the two-factor model was obtained under a precise condition of air *T* and *RH* ($T = 21°$C and $RH = 27\%$). In theory, the model should be used in these precise conditions and, for different *T* and *RH* values, it is no longer valid. US signal attenuation , which is the main source of TOF estimation errors, and the speed of sound (*s*) value are both influenced by air *T* and *RH* (Jakevicius and Demcenko 2008). In general, the effect of *RH* can be left out, especially for moderate variations (i.e., ΔRH not larger than 30–40%—condition generally satisfied within shop-floors). Moreover, if *T* variations are limited (i.e., ΔT contained within 8–10°C—condition generally satisfied within shop-floors), the effect of *T* on ultrasound attenuation can be discounted as well (Bohn 1988; Mastrogiacomo and Maisano 2010). Thus, the only effect to be compensated is that of *T* on *s*. To that purpose, *T* is periodically evaluated by embedded thermometers at the Cricket receivers and *s* is automatically updated using an experimental relation $s = s(T)$ (Franceschini et al. 2009c).

7.3 Performance Enhancing for Other Distributed Systems

Although the model presented in Sect. 7.2 has been "tailor made" for MScMS-I and the specific ultrasound-based technology, the same logic can be extended to other distributed systems with different technologies. Typically, for this kind of systems, activities in the measurement phase consist of three steps:

Table 7.5 Basic steps related to the measurement activities of three distributed systems (i.e., MScMS-I, MScMS-II and iGPS^{TM})

Distributed system	MScMS-I	MScMS-II	iGPS^{TM}
Step 1:			
Measurement of base quantities by local network devices	TOF between each network Cricket and the positioning target(s)	2D coordinates of the positioning target(s) in the view plane of each network camera	Timing measurements between signals received by positioning target(s), referring to each network transmitter
Major error sources	Use of thresholding detection method; Temperature gradients; Non punctiform dimensions of transceivers.	Misalignment between cameras and positioning target(s); Reduction in accuracy while increasing the distance between positioning target(s) and one camera.	Non perfect planarity of fan-shaped rotating laser beams; Laser beam deflections caused by temperature changes
Step 2:			
Determination of derived quantities	Distances between network Crickets and positioning target(s)	Angles (azimuth and elevation) among network cameras and positioning target(s)	Angles (azimuth and elevation) among network transmitters and positioning target(s)
Step 3:			
Optimization technique for localizing positioning target(s)	Trilateration	Triangulation	Triangulation

1. network devices perform individual local measurements of some base quantities , related to the positioning target(s);
2. these quantities are used to calculate other derived quantities related to local network devices and positioning target(s) (generally mutual distances or angles);
3. derived quantities are used to solve a global optimization problem (i.e., trilateration or triangulation) aimed at localizing positioning target(s).

Base quantities measured by local network devices are subject to systematic errors that may affect the derived quantities and the consequent localisation of positioning target(s). For the purpose of example, Table 7.5 reports the basic steps of the measurement activities as well as the most significant systematic errors for the three distributed systems analysed in this book, i.e., MScMS-I, MScMS-II and the iGPS^{TM}.

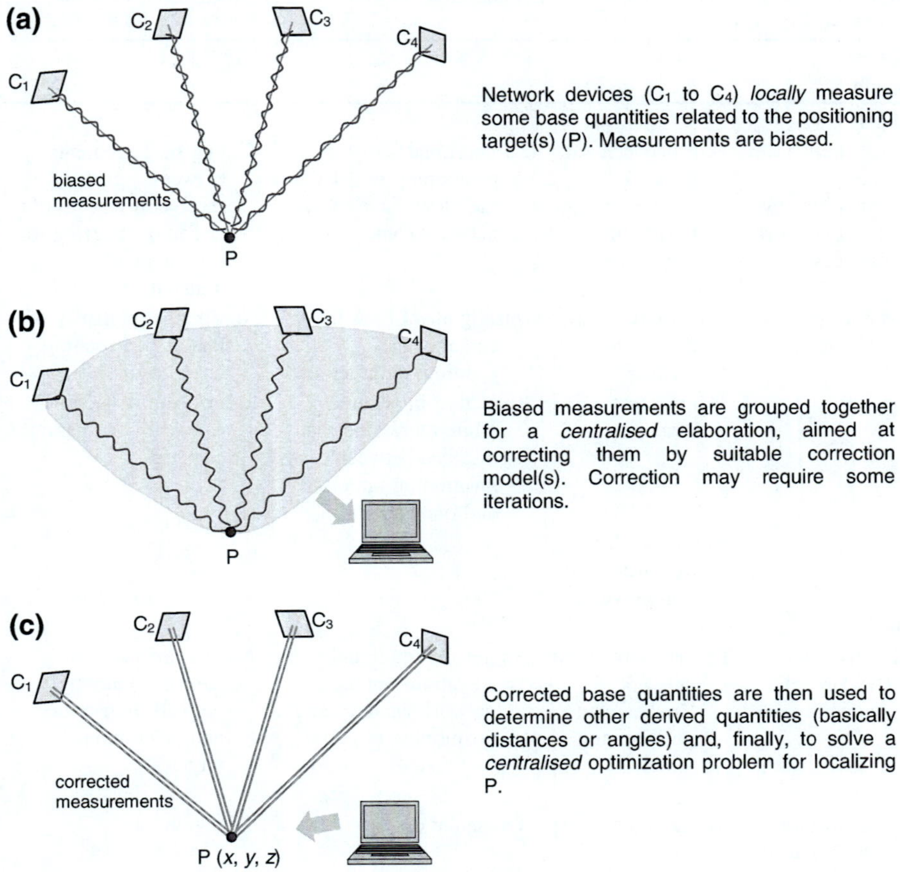

Fig. 7.21 Typical procedure of error correction for distributed measuring systems

To limit these errors, we suggest to use suitable correction models. For applying such models it is generally necessary to join together data measured by local network devices. The scheme in Fig. 7.21 illustrates the typical error correction procedure for distributed measuring systems.

For example, regarding MScMS-I it has been shown that local TOF errors can be corrected after a rough localization of the positioning target(s), which is possible only by using the TOFs measured by all the available network devices. Therefore, local base quantities are grouped together and used to run a centralised elaboration, which is aimed at correcting them. An analogous correction technique is commonly implemented by classical receivers of the satellite-based GPS, to correct systematic errors in the TOF calculation due to the effects of ionosphere, troposphere and clock synchronisation (Hofmann-Wellenhof et al. 2001).

In conclusion, we reassert that the construction of correction models is a fundamental task for enhancing metrological performance of a generic measuring instrument. To this purpose, a thorough knowledge of the technical features of the instrument as well as measurement conditions is essential. After their construction, such models should be experimentally validated in the same operational conditions in which the measuring systems will be used.

Chapter 8
Evaluation of Measurement Uncertainty

8.1 Measurement Uncertainty in LSM Distributed Systems

A crucial aspect when dealing with the problem of uncertainty evaluation of Large-Scale Dimensional Metrology distributed systems is that a well established set of reference standards is still lacking. The reason is that these systems are relatively new in comparison to other consolidated technologies, and the proposed approaches for measurement uncertainty assessment are still under discussion. At the moment, the effective reference for researchers and manufacturers is represented by the body of general metrology standards (JCGM 200:2008 2008; JCGM 100:2008 2008) and some specific standards for CMM or optical systems (Peggs et al. 2009; VDI/VDE 2634 2002).

According to the International Vocabulary of Metrology (VIM) (JCGM 200:2008 2008) and to the Guide to the Expression of Uncertainty in Measurement (GUM) (JCGM 100:2008 2008), the basic concepts related to the definition of measurement uncertainty are (see also Chap. 1):

- *Accuracy* Closeness of agreement between a measured quantity value and a true quantity value of a measurand.
- *Precision* Closeness of agreement between indications or measured quantity values obtained by replicate measurements on the same or similar objects under specified conditions (the term *precision* is also used to indicate *repeatability* and *reproducibility*).
- *Resolution* Smallest change in a quantity being measured that causes a perceptible change in the corresponding indication.
- *Error* Measured quantity value minus a reference quantity value.

There is a distinction between the two following definitions related to the concept of measurement error:

- *Systematic error* Component of measurement error that in replicate measurements remains constant or varies in a predictable manner.

F. Franceschini et al., *Distributed Large-Scale Dimensional Metrology*, DOI: 10.1007/978-0-85729-543-9_8, © Springer-Verlag London Limited 2011

- *Random error* Component of measurement error that in replicate measurements varies in an unpredictable manner.

Systematic measurement error, and its causes, can be known or unknown. A correction can be applied to compensate for a known systematic measurement error (see Chap. 7). Random errors cannot be corrected and they give the major contribution to the measurement uncertainty. Random measurement errors of a set of replicate measurements form a distribution that can be summarized by its expectation, which is generally assumed to be zero, and its variance.

Measurement uncertainty can be defined as:

- *Uncertainty* Non-negative parameter characterizing the dispersion of the quantity values being attributed to a measurand, based on the information used.

It is strictly related to the used measurement system and the implemented measuring procedure.

Putting this statement another way, uncertainty is a parameter associated with a specific measurement result that expresses the range of values about that result within which we can confidently expect to find the true value of the measurand.

A complete statement of a measurement result might look like this:

$$127.324 \pm 0.002\,\text{mm} \tag{8.1}$$

The range of values between which the true value of the measurand might lie is from 127.322 to 127.326 mm.

In Large-Scale Dimensional Metrology, when operating with distributed systems, measurement uncertainty may be evaluated according to two well distinct operating approaches.

The most typical situation is when the measurand is directly measured several times under the same conditions. In this case the measurement result is given by the arithmetic mean of the independent repeated observations and the corresponding uncertainty is gathered from the associated variance (JCGM 100:2008 2008).

Nevertheless, in most cases, the measurement value is obtained as the result of a functional relationship from two or more other quantities. This measurement is referred as indirect since the measurand is not directly measured. Uncertainty evaluation involves developing a mathematical model of random and systematic errors associated with the used instruments, the measurement procedure and the variables at stake. Hence, the overall uncertainty associated to an indirect measurement is obtained as the composition of the uncertainty of each single quantity appearing in the functional relationship. Normally, the approach used in this case is that prescribed by the Guide to the Expression of Uncertainty in Measurement (GUM) (JCGM 100:2008 2008). In substance, it is based on the Multivariate Law of Propagation of Uncertainty (MLPU).

It may be also highlighted that, in other situations, the result of a measurement is obtained by adjustment techniques. These are the cases, for example, described in Chap. 3 for the estimation of the coordinates of the Crickets mounted on the portable probe of MScMS-I (see Sect. 3.2.2), or of the probe optical markers of MScMS-II (see Sect. 3.3.3).

In this cases the task is to determine a number of unknown parameters (indirect measures) from a number of observed values (direct measures) which have a functional relationship to each other. If more observations are available than required for the determination of the unknowns, there is normally no unique solution and the unknown parameters are estimated according to functional and stochastic models. In this situation, in order to estimate the uncertainty of the unknown parameters, the Multivariate Law of Propagation of Uncertainty (MLPU) is implemented as well.

In the following Sections a brief description of most employed techniques for uncertainty evaluation is presented. For a detailed discussion readers are invited to review specialist literature (Bar-Shalom et al. 2001; JCGM 100:2008 2008).

More specifically, in Sects. 8.2–8.6 a general approach for uncertainty classification and evaluation is presented together with some simple application examples. In Sects. 8.7 and 8.8 specific applications to MScMS-I and MScMS-II are reported.

8.2 Expression of Uncertainty in Measurement

At its most basics, measurement uncertainty is a statement of how well someone thinks that they have measured something. Measurement uncertainty can, therefore, be considered a guide to the quality of the measurement. There are a number of intimidating and highly mathematical explanations of how to estimate measurement uncertainty, the ISO Guide to the Expression of Uncertainty in Measurement (GUM) (JCGM 100:2008 2008) being the definitive article.

A somewhat simpler explanation is presented here. There is no way to actually know the true value of a particular measurement because, whatever measurement process you use, there will always be some sort of error associated with the process. It follows that the result of any measurement is really only an estimate of the true value of the measurand, and so it needs to be accompanied by an estimate of the uncertainty of the correctness of the stated result.

Even after all the systematic effects have been corrected for, there remains an uncertainty due to both random effects and imperfect correction of the results of the systematic effects. The overall uncertainty estimate gives a quantitative (or numerical) assessment of the reliability of the result, and allows us to compare results with one another in a meaningful manner.

8.2.1 Type A and Type B Uncertainty

The uncertainty associated with a given estimate of a measurand can be categorized according to the method by which its numerical value is obtained.

Those uncertainties obtained by using statistical methods are termed *Type A*, whilst those obtained by methods other than statistical are termed *Type B*.

It should be noted that there is no direct correspondence between the Type A and B classifications and the old terms "random" and "systematic" (JCGM 100:2008 2008). Type A and B classifications refer to the manner in which the uncertainty was estimated, and not to the effect of the uncertainty on the measurement result. For example, the uncertainty of a correction for a known systematic effect may be obtained in some cases by a Type A evaluation while in other cases by a Type B evaluation, as well as the uncertainty characterizing a random effect.

The purpose of the Type A and Type B classification is to indicate the two different ways of evaluating uncertainty components and is for convenience of discussion only; the classification is not meant to indicate that there is any difference in the nature of the components resulting from the two types of evaluation. Both types of evaluation are based on probability distributions, and the uncertainty components resulting from either type are quantified by variances or standard deviations.

The estimated variance (denoted with u^2) characterizing an uncertainty component obtained from a Type A evaluation is calculated from series of repeated observations and is the familiar statistically estimated variance s^2. The estimated standard deviation u (i.e., the positive square root of u^2) is thus $u = s$ and for convenience is sometimes called a *Type A standard deviation* (JCGM 100:2008 2008).

For uncertainty components obtained from a Type B evaluation, the estimated variance u^2 is evaluated using the available knowledge, and the estimated standard deviation u is sometimes called a *Type B standard uncertainty*. A type B evaluation of uncertainty component is usually based on a pool of comparatively reliable information, which may include (JCGM 100:20082008):

- previous measurement data;
- experience with or general knowledge of the behaviour and properties of relevant materials and instruments;
- manufacturer's specifications;
- data provided in calibration and other certificates;
- uncertainties assigned to reference data taken from handbooks.

The proper use of the pool of available information for a Type B evaluation of standard uncertainty calls for insight based on experience and general knowledge, and is a skill that can be learned with practice. It should be recognized that a Type B evaluation of standard uncertainty can be as reliable as a Type A evaluation, especially in a measurement situation where a Type A evaluation is based on a comparatively small number of statistically independent observations (JCGM 100:2008 2008).

Thus a Type A standard uncertainty is obtained from a probability density function derived from an observed frequency distribution, while a Type B standard uncertainty is obtained from an assumed probability density function based on the degree of belief that an event will occur (often called "subjective probability") (JCGM 100:2008 2008).

8.2.2 Combined Standard Uncertainty

The *combined standard uncertainty* is defined as the standard uncertainty of the result of a measurement when the result is obtained from the values of a number of other quantities, equal to the positive square root of a sum of terms, the terms being the variance or covariances of these other quantities weighted according to how the measurement results varies with changes in these quantities (JCGM 100:2008 2008).

8.2.3 Expanded Uncertainty

The *expanded uncertainty* U, also termed as *overall uncertainty*, is defined as the quantity defining an interval about the result of a measurement that may be expected to encompass a large fraction of the distribution of values that could reasonably be attributed to the measurand. The fraction may be viewed as the coverage probability or level of confidence of an interval (JCGM 100:2008 2008).

To associate a specific level of confidence with the interval defined by the expanded uncertainty requires explicit or implicit assumptions regarding the probability distributions characterized by the measurement results and its combined standard uncertainty. The level of confidence that may be attributed to this interval can be known only to the extent to which such assumption may be justified.

8.2.4 Coverage Factor

The coverage factor k is a numerical factor used as multiplier of the standard uncertainty in order to obtain an expanded uncertainty. A coverage factor is typically in the range from 2 to 3 (JCGM 100:2008 2008). Assuming a normal distribution of measurement results, $k = 2$ corresponds to a coverage probability equal to 95%, while $k = 3$ corresponds to 99.7% (see Fig. 8.1).

8.3 Uncertainty Evaluation with Independent Repeated Observations

When a direct measurement (i.e., the measurand is measured directly) can be repeated n times under unchanged measurement conditions, it can be treated as a random variable X. In this case, under the assumption of absence of systematic effects, the best estimate of its expectation or expected value, under the hypothesis of independent observations, is the arithmetic mean $\bar{x}$ (or average) of the observations x_i ($i = 1 \ldots n$):

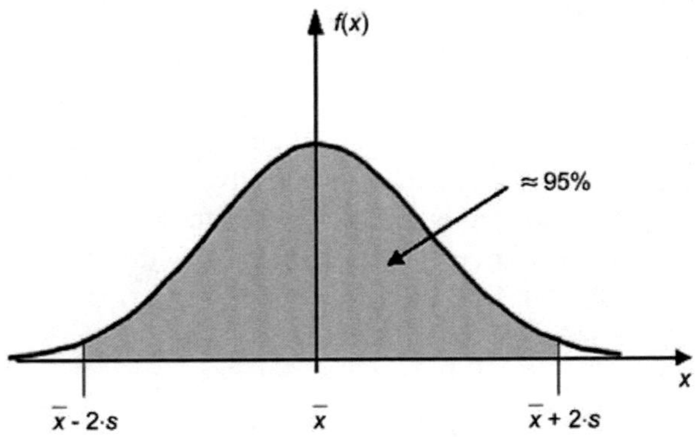

Fig. 8.1 Coverage probability (*gray*), i.e., the fraction of the distribution of measured values that could reasonably be attributed to the measurand, corresponding to a coverage factor $k = 2$, when assuming a normal distribution

$$\bar{x} = \frac{1}{n}\sum_{i=1}^{n} x_i \tag{8.2}$$

The individual observations x_i differ in value because of random variations in the influence quantities (random effects). The statistically estimated variance of the observations is given by:

$$s^2 = \frac{1}{n-1}\sum_{i=1}^{n} (x_i - \bar{x})^2 \tag{8.3}$$

This estimate of variance and its positive square root s (experimental standard deviation of the single value) characterizes the variability of the observed values x_i, or more specifically, their dispersion about their mean $\bar{x}$.

Furthermore, the experimental standard deviation of the mean is:

$$s_{\bar{x}} = \frac{s}{\sqrt{n}} \tag{8.4}$$

The experimental standard deviation of the single value s and the experimental standard deviation of the mean $s_{\bar{x}}$ may be used as a "measure" of the uncertainty of the measurement (Type A uncertainty).

Hence, the standard uncertainty of the single value and of the mean are respectively $u = s$ and $u_{\bar{x}} = s_{\bar{x}}$.

Usually the so called "expanded uncertainty" is used in order to refer to an uncertainty interval expected to encompass a given fraction of values that could reasonably be attributed to the measurand. The expanded uncertainty is obtained by multiplying the standard uncertainty by the coverage factor k, which in most of cases is fixed at $k = 2$ (that means that, assuming a normal distribution of measurement results, the corresponding coverage probability is equal to 95%).

In particular, the expanded uncertainty of a single measure is defined as:

$$U_x = 2 \cdot u \tag{8.5}$$

while the expanded uncertainty of the mean of n observations is:

Table 8.1 Example of 40 measurements of the length of a steel bar

Measurements (mm)							
127.81	127.10	127.93	127.67	127.27	127.39	127.90	127.81
127.70	127.51	127.29	128.04	127.45	127.08	127.52	127.70
127.13	127.81	127.42	127.32	127.97	127.22	127.93	127.13
127.90	127.20	127.28	127.61	127.54	127.10	127.57	127.90
127.51	127.06	127.97	127.64	127.83	127.85	127.06	127.51

$$U_{\bar{x}} = 2 \cdot u_{\bar{x}} \tag{8.6}$$

Hence, according to the prescriptions of the Guide to the Expression of Uncertainty in Measurement (GUM) (JCGM 100:2008 2008), the result of n independent repeated measurements (under the same measurement conditions) can be expressed as follows:

$$\bar{x} \pm 2 \cdot u \tag{8.7}$$

if referring to a single measurement, or

$$\bar{x} \pm 2 \cdot u_{\bar{x}} \tag{8.8}$$

if referring to their mean.

Consider, for example, a set of 40 measurements of the length of a steel bar, reported in Table 8.1.

The corresponding average value is $\bar{x} = 127.51$ mm, and the estimate of the standard deviation is $s = 0.31$ mm. Hence, assuming a coverage factor equal to 2, the resulting expanded uncertainty of the single value and of the mean are $U_x = k{\cdot}s_x = 2{\cdot}0.31 = 0.62$ mm and $U_{\bar{x}} = k \cdot \frac{s_{\bar{x}}}{\sqrt{n}} = 2 \cdot \frac{0.31}{\sqrt{40}} = 0.10$ mm, respectively.

8.4 Evaluation of Combined Uncertainty

When a measurand Y is not directly measured, but is determined from N other quantities X_1, X_2, …, X_N through a functional relationship f:

$$Y = f(X_1, X_2, \ldots, X_N) \tag{8.9}$$

the corresponding uncertainty may be estimated using the Multivariate Law of Propagation of Uncertainty (MLPU) MLPU (JCGM 100:2008 2008).

In the most general case this entails a multidimensional response, hence the measurement result can be expressed as a vector $\boldsymbol{Y} = [Y_1 \ldots Y_M]^T$ function of a vector of estimates $\boldsymbol{X} = [X_1 \ldots X_N]^T$:

$$\boldsymbol{Y} = \boldsymbol{f}(\boldsymbol{X}) \tag{8.10}$$

Consider for example a set of $M = 5$ distances from a given Cricket (see Chap. 3) to 5 network devices. Suppose each distance has been obtained by measuring the corresponding TDoA and applying Eq. 3.8 with an additive term for the correction of distance error due to the relative orientations between the network device and the Cricket (see Chap. 7). The vector of the five distances can be expressed as:

$$\boldsymbol{Y} = \begin{bmatrix} d_1 \\ d_2 \\ d_3 \\ d_4 \\ d_5 \end{bmatrix} = \begin{bmatrix} s \cdot \Delta t_1 + C_1 \\ s \cdot \Delta t_2 + C_2 \\ s \cdot \Delta t_3 + C_3 \\ s \cdot \Delta t_4 + C_4 \\ s \cdot \Delta t_5 + C_5 \end{bmatrix} \tag{8.11}$$

where s is the speed of sound (about 340 m/s in air, with temperature $T = 20°\text{C}$ and relative humidity RH = 50%), Δt_i is the TDoA corresponding to the i-th network device ($i = 1, \ldots, 5$) and C_i is the related correction term.

Vector $\boldsymbol{Y}$ can, hence, be expressed as in Eq. 8.10 by defining:

$$\boldsymbol{Y} = \boldsymbol{f}(\boldsymbol{X}) = \begin{bmatrix} f_1(\boldsymbol{X}) \\ f_2(\boldsymbol{X}) \\ f_3(\boldsymbol{X}) \\ f_4(\boldsymbol{X}) \\ f_5(\boldsymbol{X}) \end{bmatrix} = \begin{bmatrix} s \cdot \Delta t_1 + C_1 \\ s \cdot \Delta t_2 + C_2 \\ s \cdot \Delta t_3 + C_3 \\ s \cdot \Delta t_4 + C_4 \\ s \cdot \Delta t_5 + C_5 \end{bmatrix} \tag{8.12}$$

where

$$\begin{aligned} \boldsymbol{X} &= [X_1 \quad X_2 \quad X_3 \quad X_4 \quad X_5 \quad X_6 \quad X_7 \quad X_8 \quad X_9 \quad X_{10}]^T \\ &= [\Delta t_1 \quad \Delta t_2 \quad \Delta t_3 \quad \Delta t_4 \quad \Delta t_5 \quad C_1 \quad C_2 \quad C_3 \quad C_4 \quad C_5]^T \end{aligned} \tag{8.13}$$

If each component of vector $\boldsymbol{f}(\boldsymbol{X})$ in Eq. 8.10 is expanded into a Taylor series around the average values $\bar{\boldsymbol{x}}$ of the estimates vector $\boldsymbol{X}$, ignoring terms after the first order, it can be demonstrated that an estimate of covariance matrix $\Sigma_Y \in \mathbb{R}^{M,M}$ associated with $\boldsymbol{Y}$ is given by (JCGM 100:2008 2008):

$$\hat{\Sigma}_Y = \boldsymbol{J}\Sigma_X\boldsymbol{J}^T \tag{8.14}$$

where $\boldsymbol{J} \in \mathbb{R}^{M,N}$ is the Jacobian (or sensitivity matrix) of $\boldsymbol{f}(\boldsymbol{X})$:

$$J_{i,j} = \left.\frac{\partial Y_i}{\partial X_j}\right|_{\bar{x}} \quad i = 1, \ldots, M \; j = 1, \ldots, N \tag{8.15}$$

and $\Sigma_X \in \mathbb{R}^{N,N}$ is the covariance matrix of X.

If we consider again the previous example of a set of $M = 5$ distances from a given Cricket to 5 network devices, the resulting Jacobian $\left(J \in \mathbb{R}^{5,10}\right)$ is:

$$
\boldsymbol{J} = \begin{bmatrix} \left.\frac{\partial f_1(\boldsymbol{X})}{\partial \Delta t_1}\right|_{\bar{x}} & \left.\frac{\partial f_1(\boldsymbol{X})}{\partial \Delta t_2}\right|_{\bar{x}} & \cdots & \left.\frac{\partial f_1(\boldsymbol{X})}{\partial C_5}\right|_{\bar{x}} \\ \left.\frac{\partial f_2(\boldsymbol{X})}{\partial \Delta t_1}\right|_{\bar{x}} & \left.\frac{\partial f_2(\boldsymbol{X})}{\partial \Delta t_2}\right|_{\bar{x}} & \cdots & \left.\frac{\partial f_2(\boldsymbol{X})}{\partial C_5}\right|_{\bar{x}} \\ \vdots & \vdots & \ddots & \vdots \\ \left.\frac{\partial f_5(\boldsymbol{X})}{\partial \Delta t_1}\right|_{\bar{x}} & \left.\frac{\partial f_5(\boldsymbol{X})}{\partial \Delta t_2}\right|_{\bar{x}} & \cdots & \left.\frac{\partial f_5(\boldsymbol{X})}{\partial C_5}\right|_{\bar{x}} \end{bmatrix} = \begin{bmatrix} s & 0 & 0 & 0 & 0 & 1 & 0 & 0 & 0 & 0 \\ 0 & s & 0 & 0 & 0 & 0 & 1 & 0 & 0 & 0 \\ 0 & 0 & s & 0 & 0 & 0 & 0 & 1 & 0 & 0 \\ 0 & 0 & 0 & s & 0 & 0 & 0 & 0 & 1 & 0 \\ 0 & 0 & 0 & 0 & s & 0 & 0 & 0 & 0 & 1 \end{bmatrix} \tag{8.16}
$$

and the corresponding estimate of $\Sigma_X \in \mathbb{R}^{10,10}$ matrix is:

$$
\hat{\Sigma}_X = \begin{bmatrix} u^2(X_1) & u(X_1, X_2) & \cdots & u(X_1, X_{10}) \\ u(X_2, X_1) & u^2(X_2) & \cdots & u(X_2, X_{10}) \\ \vdots & \vdots & \ddots & \vdots \\ u(X_{10}, X_1) & u(X_{10}, X_2) & \cdots & u^2(X_{10}) \end{bmatrix} = \begin{bmatrix} u^2(\Delta t_1) & u(\Delta t_1, \Delta t_2) & \cdots & u(\Delta t_1, C_5) \\ u(\Delta t_2, \Delta t_1) & u^2(\Delta t_2) & \cdots & u(\Delta t_2, C_5) \\ \vdots & \vdots & \ddots & \vdots \\ u(C_5, \Delta t_1) & u(C_5, \Delta t_2) & \cdots & u^2(C_5) \end{bmatrix} \tag{8.17}
$$

Hence the components of $\hat{\Sigma}_Y \in \mathbb{R}^{5,5}$ matrix may be obtained by Eq. 8.14. For the sake of example, the first component $\left(\hat{\Sigma}_{Y,1,1}\right)$ is reported hereafter:

$$
\begin{aligned} \hat{\Sigma}_{Y,1,1} &= \sum_{i=1}^{10}\sum_{l=1}^{10} \left.\frac{\partial f_1(\boldsymbol{X})}{\partial X_i}\right|_{\bar{x}} \left.\frac{\partial f_1(\boldsymbol{X})}{\partial X_l}\right|_{\bar{x}} u(x_i, x_l) \\ &= \sum_{i=1}^{10} \left(\left.\frac{\partial f_1(\boldsymbol{X})}{\partial X_i}\right|_{\bar{x}} \right)^2 u^2(x_i) + 2\sum_{i=1}^{9}\sum_{l=i+1}^{10} \left.\frac{\partial f_1(\boldsymbol{X})}{\partial X_i}\right|_{\bar{x}} \left.\frac{\partial f_1(\boldsymbol{X})}{\partial X_l}\right|_{\bar{x}} u(x_i, x_l) \end{aligned} \tag{8.18}
$$

The same holds for other components of $\hat{\Sigma}_Y$.

This method can be applied only if the covariance matrix Σ_X is a priori known. That means that it should be estimated by a preliminary experimental session, or hypothesized on the basis of the previous knowledge of the phenomenon under study. In the case of independent observations, the covariance terms become zero and the covariance matrix Σ_X is reduced to a diagonal matrix. This happens in many uncertainty propagation problems where either independent observations are given, or no significant knowledge about correlations between observations is available.

The obtained $\hat{\Sigma}_Y$ matrix contains the estimates of both the variance (matrix diagonal elements) and covariance (matrix elements out of the diagonal) terms of the elements of vector $\boldsymbol{Y}$.

Specifically, the extended uncertainty (with coverage factor $k = 2$) of each element of $\boldsymbol{Y}$ can be obtained from the diagonal elements of matrix Σ_Y as follows:

$$U_{Y,i} = 2 \cdot \sqrt{\hat{\Sigma}_{Y,i,i}} \tag{8.19}$$

Referring to the above example, the resulting covariance matrix $\hat{\Sigma}_Y \in \mathbb{R}^{5,5}$, in absence of covariance is:

$$\hat{\Sigma}_Y = \begin{bmatrix} \hat{\Sigma}_{Y,1,1} & 0 & 0 & 0 & 0 \\ 0 & \hat{\Sigma}_{Y,2,2} & 0 & 0 & 0 \\ 0 & 0 & \hat{\Sigma}_{Y,3,3} & 0 & 0 \\ 0 & 0 & 0 & \hat{\Sigma}_{Y,4,4} & 0 \\ 0 & 0 & 0 & 0 & \hat{\Sigma}_{Y,5,5} \end{bmatrix} \tag{8.20}$$

where:

$$\begin{aligned} \hat{\Sigma}_{Y,1,1} &= \sum_{i=1}^{10} \left(\left. \frac{\partial f_1(\boldsymbol{X})}{\partial X_i} \right|_{\bar{x}} \right)^2 u^2(x_i) \\ \hat{\Sigma}_{Y,2,2} &= \sum_{i=1}^{10} \left(\left. \frac{\partial f_2(\boldsymbol{X})}{\partial X_i} \right|_{\bar{x}} \right)^2 u^2(x_i) \\ \hat{\Sigma}_{Y,3,3} &= \sum_{i=1}^{10} \left(\left. \frac{\partial f_3(\boldsymbol{X})}{\partial X_i} \right|_{\bar{x}} \right)^2 u^2(x_i) \\ \hat{\Sigma}_{Y,4,4} &= \sum_{i=1}^{10} \left(\left. \frac{\partial f_4(\boldsymbol{X})}{\partial X_i} \right|_{\bar{x}} \right)^2 u^2(x_i) \\ \hat{\Sigma}_{Y,5,5} &= \sum_{i=1}^{10} \left(\left. \frac{\partial f_5(\boldsymbol{X})}{\partial X_i} \right|_{\bar{x}} \right)^2 u^2(x_i) \end{aligned} \tag{8.21}$$

The great potentiality of the Multivariate Law of Propagation of Uncertainty (MLPU) is that, in absence of systematic errors, it can be applied before the measurement is performed, as well as after it. This is very helpful during the phases of design of the measurement procedure. If Σ_X is a priori known, the measurement set-up configuration and procedure may be adequately arranged (that corresponds to modify the terms of $\boldsymbol{J}$ matrix) in order to minimize the terms of the output matrix $\hat{\Sigma}_Y$.

8.5 Least-Squares Adjustment

The approach for uncertainty evaluation is different when referring to a set of unknown parameters determined starting from an overabundant set of different measurements (i.e., the number of measurement is greater than the number of parameters) of an observable variable associated to the parameters by a functional relationship. In this case the values of the unknown parameters are estimated by applying specific procedures, known as "adjustment techniques". Consider, for example, a set of distances from a given Cricket (see Chap. 3), fixed in an unknown position, to the devices of a network. The distances are directly measured by TDoA (Time Difference of Arrival) technique. The coordinates of each network device are a priori known. The goal is to determine the 3D coordinates of the given Cricket. The functional relationship between the measured distances and the unknown parameters (3D coordinates of the Cricket) is given by the Pythagorean theorem. In this case, the three unknown parameters may be estimated using a particular adjustment technique called "least-squares method" if at least four distance measurements have been performed.

More in general, if we consider M observations (measured values), acquired in different measurement conditions, they can be rewritten in a vector $\boldsymbol{Y}$ ("observation vector") as follows:

$$\boldsymbol{Y} = [Y_1 \ldots Y_M]^T \tag{8.22}$$

Since the elements of the observation vector are measured data, they are affected by measurement uncertainty which produces small random variation between one observation and the other. It is assumed that these small variations follow a normal distribution with zero mean and a fixed standard deviation. Furthermore each measurement is free of systematic errors.

The N unknown parameters form the vector of unknowns $\boldsymbol{X}$, also called "parameter vector":

$$\boldsymbol{X} = [X_1 \ldots X_N]^T \tag{8.23}$$

The number of observations is assumed to be greater than the number of unknowns ($M > N$).

The functional model describes the relationship between the "true" observation values $\tilde{\boldsymbol{Y}}$ and the "true" values of the unknowns $\tilde{\boldsymbol{X}}$. This relationship is expressed by a vector of functions φ of the unknowns:

$$\tilde{\boldsymbol{Y}} = \varphi(\tilde{\boldsymbol{X}}) = \left[\varphi_1(\tilde{\boldsymbol{X}}) \ldots \varphi_M(\tilde{\boldsymbol{X}})\right]^T \tag{8.24}$$

Since the true values are normally not known, the observation vector $\tilde{\boldsymbol{Y}}$ is replaced by the estimated observation vector $\hat{\boldsymbol{Y}}$, i.e., by the vector $\boldsymbol{Y}$ of the measured observations and the associated vector of small residuals $\boldsymbol{v}$. Similarly, the vector of unknowns is replaced by the estimated (adjusted) unknowns $\hat{\boldsymbol{X}}$. As a result, the following non-linear correction equations are obtained:

$$\hat{\boldsymbol{Y}} = \boldsymbol{Y} + \boldsymbol{v} = \varphi(\hat{\boldsymbol{X}}) \tag{8.25}$$

If approximate values $\boldsymbol{X^0}$ of the unknowns are available, the vector of unknowns can be expressed as the following sum:

$$\hat{\boldsymbol{X}} = \boldsymbol{X^0} + \Delta\hat{\boldsymbol{X}} \tag{8.26}$$

where only the $\Delta\hat{\boldsymbol{X}}$ values must be determined (consider that in most cases $\boldsymbol{X}^0$ is fixed equal to $\boldsymbol{0}$).

According to the approximate values of $\boldsymbol{X}^0$, approximate values of the observations can be calculated using the functional model:

$$\boldsymbol{Y}^0 = \varphi(\boldsymbol{X}^0) \tag{8.27}$$

In this way, reduced observations (observed minus calculated) are obtained:

$$\Delta\boldsymbol{Y} = \boldsymbol{Y} - \boldsymbol{Y}^0 \tag{8.28}$$

For sufficiently small values of $\Delta\hat{\boldsymbol{X}}$, the correction equations (Eq. 8.25) can be expanded into a Taylor series around the approximate values $\boldsymbol{X}^0$, ignoring terms after the first order:

$$\boldsymbol{Y} + \boldsymbol{v} = \varphi(\boldsymbol{X}^0) + \left(\frac{\partial\varphi(\boldsymbol{X})}{\partial\boldsymbol{X}}\right)_0 (\hat{\boldsymbol{X}} - \boldsymbol{X}^0) = \boldsymbol{Y}^0 + \left(\frac{\partial\varphi(\boldsymbol{X})}{\partial\boldsymbol{X}}\right)_0 \Delta\hat{\boldsymbol{X}} \tag{8.29}$$

where $\left(\frac{\partial\varphi(\boldsymbol{X})}{\partial\boldsymbol{X}}\right)_0$ is the Jacobian matrix calculated in $\boldsymbol{X}^0$ (also called "design matrix" $\boldsymbol{A} \in \mathbb{R}^{M,N}$):

$$\boldsymbol{A} = \left(\frac{\partial\varphi(\boldsymbol{X})}{\partial\boldsymbol{X}}\right)_0 = \begin{bmatrix} \left(\frac{\partial\varphi_1(\boldsymbol{X})}{\partial X_1}\right)_0 & \left(\frac{\partial\varphi_1(\boldsymbol{X})}{\partial X_2}\right)_0 & \cdots & \left(\frac{\partial\varphi_1(\boldsymbol{X})}{\partial X_N}\right)_0 \\ \left(\frac{\partial\varphi_2(\boldsymbol{X})}{\partial X_1}\right)_0 & \left(\frac{\partial\varphi_2(\boldsymbol{X})}{\partial X_2}\right)_0 & \cdots & \left(\frac{\partial\varphi_2(\boldsymbol{X})}{\partial X_N}\right)_0 \\ \vdots & \vdots & \ddots & \vdots \\ \left(\frac{\partial\varphi_M(\boldsymbol{X})}{\partial X_1}\right)_0 & \left(\frac{\partial\varphi_M(\boldsymbol{X})}{\partial X_2}\right)_0 & \cdots & \left(\frac{\partial\varphi_M(\boldsymbol{X})}{\partial X_N}\right)_0 \end{bmatrix} \tag{8.30}$$

Hence, according to Eq. 8.29, the linearized correction equations are:

$$\Delta\tilde{\boldsymbol{Y}} = \Delta\boldsymbol{Y} + \boldsymbol{v} = \boldsymbol{A}\Delta\hat{\boldsymbol{X}} \tag{8.31}$$

The stochastic properties of the observations $\boldsymbol{Y}$ are defined by the covariance matrix $\Sigma_Y \in \mathbb{R}^{M,M}$:

$$\Sigma_Y = \begin{bmatrix} \sigma_1^2 & \rho_{12}\sigma_1\sigma_2 & \cdots & \rho_{1M}\sigma_1\sigma_M \\ \rho_{21}\sigma_2\sigma_1 & \sigma_2^2 & \cdots & \rho_{2M}\sigma_2\sigma_M \\ \vdots & \vdots & \ddots & \vdots \\ \rho_{M1}\sigma_M\sigma_1 & \rho_{M2}\sigma_M\sigma_2 & \cdots & \sigma_M^2 \end{bmatrix} \tag{8.32}$$

where σ_i is the standard deviation of observation Y_i ($i = 1, \ldots, M$), and ρ_{ij} is the correlation coefficient between Y_i and Y_j ($i \neq j$).

Introducing the multiplication factor σ_0^2, the cofactor matrix $\boldsymbol{Q}_Y \in \mathbb{R}^{M,M}$ of observations is obtained:

$$\boldsymbol{Q}_Y = \frac{1}{\sigma_0^2}\Sigma_Y \tag{8.33}$$

The inverse matrix of Q_Y is the weight matrix:

$$\boldsymbol{P}_Y = \boldsymbol{Q}_Y^{-1} = \sigma_0^2\Sigma_Y^{-1} \tag{8.34}$$

The covariance matrix Σ_Y is the only component containing information about the efficiency of the functional model in the adjustment process. In the case of independent observations, such as for the propagation law, the correlation coefficients ρ_{ij}, with $i \neq j$, become zero and the covariance matrix is reduced to a diagonal matrix. This is the standard case for many adjustment problems where either independent observations are given, or no significant knowledge about correlations between observations is available.

In this case the weight matrix assumes the following form:

$$\boldsymbol{P}_Y = \begin{bmatrix} \frac{\sigma_0^2}{\sigma_1^2} & 0 & \cdots & 0 \\ 0 & \frac{\sigma_0^2}{\sigma_2^2} & \cdots & 0 \\ \vdots & \vdots & \ddots & \vdots \\ 0 & 0 & \cdots & \frac{\sigma_0^2}{\sigma_M^2} \end{bmatrix} = \begin{bmatrix} p_1 & 0 & \cdots & 0 \\ 0 & p_2 & \cdots & 0 \\ \vdots & \vdots & \ddots & \vdots \\ 0 & 0 & \cdots & p_M \end{bmatrix} \tag{8.35}$$

If the following equality holds:

$$\sigma_i = \sigma_0 \tag{8.36}$$

for $i = 1, \ldots, M$, $\boldsymbol{P}_Y$ becomes the identity matrix $\boldsymbol{I}$. σ_0 is the true value of the standard deviation of unit weight (standard deviation of an observation with weight 1). It can be regarded as a multiplication constant.

Usually the true standard deviation σ_0 is not known in practical applications and the empirical standard deviation s_0 is used instead. Here s_0 denotes the standard deviation a priori , while $\tilde{s}_0$ represents the standard deviation a posteriori ("adjusted standard deviation"), obtained using the residuals of the adjustment. The empirical standard deviation is only meaningful in cases of significant redundancy of measurements.

Even if in literature many different adjustment techniques are proposed, the most used in localization approaches is the least-squares method, which is based on the idea that the unknown parameters are estimated with maximum probability. This method ensures that the unknowns are estimated unbiased and with minimum variance. The working principle consists in identifying those parameters which guarantee the minimum value of the sum of the squared residuals (elements of vector $\boldsymbol{v}$).

Starting from Eq. 8.31 the following expression may be obtained ("normal equations"):

$$\boldsymbol{N}\Delta\hat{\boldsymbol{X}} - \boldsymbol{T} = \boldsymbol{0} \tag{8.37}$$

where $\boldsymbol{N} \in \mathbb{R}^{M,M}$ is the matrix of normal equations

$$\boldsymbol{N} = \boldsymbol{A}^T\boldsymbol{P}_Y\boldsymbol{A} \tag{8.38}$$

and $\boldsymbol{T} \in \mathbb{R}^{M,1}$ is the absolute term vector:

$$\boldsymbol{T} = \boldsymbol{A}^T\boldsymbol{P}_Y\Delta\boldsymbol{Y} \tag{8.39}$$

The solution of Eq. 8.37 produces the vector of estimates (Bar-Shalom et al. 2001):

$$\Delta\hat{\boldsymbol{X}} = \left(\boldsymbol{A}^T\boldsymbol{P}_Y\boldsymbol{A}\right)^{-1}\left(\boldsymbol{A}^T\boldsymbol{P}_Y\Delta\boldsymbol{Y}\right) \tag{8.40}$$

According to Eq. 8.26, the vector of the estimates of the unknowns is given by:

$$\hat{\boldsymbol{X}} = \boldsymbol{X}^0 + \Delta\hat{\boldsymbol{X}} \tag{8.41}$$

and the corresponding cofactor matrix of unknowns may be defined as:

$$\boldsymbol{Q}_X = \boldsymbol{N}^{-1} = \left(\boldsymbol{A}^T\boldsymbol{P}_Y\boldsymbol{A}\right)^{-1} \tag{8.42}$$

Considering that, according to Eq. 8.31, the residuals may be calculated as:

$$\boldsymbol{v} = \boldsymbol{A}\Delta\hat{\boldsymbol{X}} - \Delta\boldsymbol{Y} \tag{8.43}$$

the standard deviation a posteriori results:

$$\tilde{s}_0 = \sqrt{\frac{\boldsymbol{v}^T\boldsymbol{P}_Y\boldsymbol{v}}{M - N}} \tag{8.44}$$

and the estimated covariance matrix of unknowns:

$$\hat{\Sigma}_X = \tilde{s}_0^2\boldsymbol{Q}_X \tag{8.45}$$

The Eq. 8.40 must be solved iteratively if only coarse approximate values are given for non-linear problems. In this case the corrected approximate values in the l-th iteration are used as starting values for the linearized functional model of the $(l + 1)$-th iteration, until the sum of added corrections for the unknowns is less than a given threshold.

In order to better understand the least-squares adjustment procedure, consider the $M = 6$ distances reported in Table 8.2. They have been measured from a Cricket (see Sect. 3.2.2) positioned inside a measurement space and 6 network devices positioned all around it. The 3D coordinates of each network device are known (see Table 8.2). The objective is to determine the 3D coordinates of the Cricket.

Table 8.2 Example of 6 distances measured from a Cricket (see Sect. 3.2.2) positioned inside a measurement space and 6 network devices positioned all around it

	Measured distance d_i (m)	x_i (m)	y_i (m)	z_i (m)
Device 1	2.454	1.500	1.500	2.500
Device 2	3.010	1.500	2.500	2.500
Device 3	3.005	2.500	1.500	2.500
Device 4	3.474	2.500	2.500	2.500
Device 5	3.749	3.500	1.500	2.500
Device 6	4.127	3.500	2.500	2.500

The functional relationship between distances (measured observations), coordinates of network devices and Cricket coordinates (unknowns) is given by the Pythagorean theorem:

$$d_i = \sqrt{(x_i - x_P)^2 + (y_i - y_P)^2 + (y_i - y_P)^2} \tag{8.46}$$

where d_i (with $i = 1, \ldots, 6$) are the measured distances, (x_i, y_i, z_i) (with $i = 1, \ldots, 6$) are the known 3D coordinates of the 6 network devices, and (x_P, y_P, z_P) are the unknown 3D coordinates of the Cricket.

In order to give approximate values of the unknowns (expressed in meters), we suppose that the Cricket is positioned in the origin of the axes, hence

$$\begin{bmatrix} x_P^0 & y_P^0 & z_P^0 \end{bmatrix}^T = \begin{bmatrix} 0 & 0 & 0 \end{bmatrix}^T = \mathbf{0}^T. \tag{8.47}$$

The resulting vector of reduced observations $\Delta\boldsymbol{d}$ (expressed in meters) is (Eq. 8.28):

$$\Delta\boldsymbol{d} = \begin{bmatrix} d_1 \\ d_2 \\ d_3 \\ d_4 \\ d_5 \\ d_6 \end{bmatrix} - \begin{bmatrix} \sqrt{(x_1 - 0)^2 + (y_1 - 0)^2 + (z_1 - 0)^2} \\ \sqrt{(x_2 - 0)^2 + (y_2 - 0)^2 + (z_2 - 0)^2} \\ \sqrt{(x_3 - 0)^2 + (y_3 - 0)^2 + (z_3 - 0)^2} \\ \sqrt{(x_4 - 0)^2 + (y_4 - 0)^2 + (z_4 - 0)^2} \\ \sqrt{(x_5 - 0)^2 + (y_5 - 0)^2 + (z_5 - 0)^2} \\ \sqrt{(x_6 - 0)^2 + (y_6 - 0)^2 + (z_6 - 0)^2} \end{bmatrix}$$

$$= \begin{bmatrix} 2.454 \\ 3.010 \\ 3.005 \\ 3.474 \\ 3.749 \\ 4.127 \end{bmatrix} - \begin{bmatrix} 3.279 \\ 3.841 \\ 3.841 \\ 4.330 \\ 4.555 \\ 4.975 \end{bmatrix} = \begin{bmatrix} -0.825 \\ -0.831 \\ -0.836 \\ -0.856 \\ -0.806 \\ -0.848 \end{bmatrix} \tag{8.48}$$

The related design matrix is (Eq. 8.30):

$$A = \begin{bmatrix} \frac{-x_1}{\sqrt{(x_1)^2+(y_1)^2+(z_1)^2}} & \frac{-y_1}{\sqrt{(x_1)^2+(y_1)^2+(z_1)^2}} & \frac{-z_1}{\sqrt{(x_1)^2+(y_1)^2+(z_1)^2}} \\ \frac{-x_2}{\sqrt{(x_2)^2+(y_2)^2+(z_2)^2}} & \frac{-y_2}{\sqrt{(x_2)^2+(y_2)^2+(z_2)^2}} & \frac{-z_2}{\sqrt{(x_2)^2+(y_2)^2+(z_2)^2}} \\ \frac{-x_3}{\sqrt{(x_3)^2+(y_3)^2+(z_3)^2}} & \frac{-y_3}{\sqrt{(x_3)^2+(y_3)^2+(z_3)^2}} & \frac{-z_3}{\sqrt{(x_3)^2+(y_3)^2+(z_3)^2}} \\ \frac{-x_4}{\sqrt{(x_4)^2+(y_4)^2+(z_4)^2}} & \frac{-y_4}{\sqrt{(x_4)^2+(y_4)^2+(z_4)^2}} & \frac{-z_4}{\sqrt{(x_4)^2+(y_4)^2+(z_4)^2}} \\ \frac{-x_5}{\sqrt{(x_5)^2+(y_5)^2+(z_5)^2}} & \frac{-y_5}{\sqrt{(x_5)^2+(y_5)^2+(z_5)^2}} & \frac{-z_5}{\sqrt{(x_5)^2+(y_5)^2+(z_5)^2}} \\ \frac{-x_6}{\sqrt{(x_6)^2+(y_6)^2+(z_6)^2}} & \frac{-y_6}{\sqrt{(x_6)^2+(y_6)^2+(z_6)^2}} & \frac{-z_6}{\sqrt{(x_6)^2+(y_6)^2+(z_6)^2}} \end{bmatrix}$$

$$= \begin{bmatrix} -0.45750 & -0.45750 & -0.76249 \\ -0.39057 & -0.65094 & -0.65094 \\ -0.65094 & -0.39057 & -0.65094 \\ -0.57735 & -0.57735 & -0.57735 \\ -0.76835 & -0.32929 & -0.54882 \\ -0.70353 & -0.50252 & -0.50252 \end{bmatrix} \tag{8.49}$$

Assuming no covariance between the measured distances and the same variance for each of them (i.e., the weight matrix is given by the identity matrix $\mathbf{I}$), the normal equation (Eq. 8.37) becomes:

$$\boldsymbol{A}^T\boldsymbol{IA}\begin{bmatrix} x_P - x_P^0 \\ y_P - y_P^0 \\ z_P - z_P^0 \end{bmatrix} - \boldsymbol{A}^T\boldsymbol{I}\Delta\boldsymbol{d} = 0 \tag{8.50}$$

hence:

$$\begin{bmatrix} 2.20422 & 1.65766 & 2.13536 \\ 1.65766 & 1.47987 & 1.79338 \\ 2.13536 & 1.79338 & 2.31592 \end{bmatrix}\begin{bmatrix} x_P \\ y_P \\ z_P \end{bmatrix} - \begin{bmatrix} 2.95590 \\ 2.43018 \\ 3.07627 \end{bmatrix} = \begin{bmatrix} 0 \\ 0 \\ 0 \end{bmatrix} \tag{8.51}$$

Its solution produces the estimated values (expressed in meters) of the unknowns (3D coordinates of the Cricket) (Eqs. 8.40 and 8.41):

$$[\hat{x}_P \quad \hat{y}_P \quad \hat{z}_P]^T = [0.4989 \quad 0.5044 \quad 0.4778]^T \tag{8.52}$$

Considering that the residuals (expressed in meters) are (Eq. 8.43):

$$\boldsymbol{v} = \begin{bmatrix} -0.45750 & -0.45750 & -0.76249 \\ -0.39057 & -0.65094 & -0.65094 \\ -0.65094 & -0.39057 & -0.65094 \\ -0.57735 & -0.57735 & -0.57735 \\ -0.76835 & -0.32929 & -0.54882 \\ -0.70353 & -0.50252 & -0.50252 \end{bmatrix}\begin{bmatrix} 0.4989 \\ 0.5044 \\ 0.4778 \end{bmatrix} - \begin{bmatrix} -0.825 \\ -0.831 \\ -0.836 \\ -0.856 \\ -0.806 \\ -0.848 \end{bmatrix} = \begin{bmatrix} 0.0014 \\ -0.0036 \\ 0.0028 \\ 0.0011 \\ -0.0054 \\ 0.0034 \end{bmatrix} \tag{8.53}$$

the associated standard deviation a posteriori (expressed in meters) is (Eq. 8.44):

$$\tilde{s}_0 = \sqrt{\frac{1}{6-3}\begin{bmatrix} 0.0014 \\ -0.0036 \\ 0.0028 \\ 0.0011 \\ -0.0054 \\ 0.0034 \end{bmatrix}\begin{bmatrix} 0.0014 \\ -0.0036 \\ 0.0028 \\ 0.0011 \\ -0.0054 \\ 0.0034 \end{bmatrix}^T} = 0.0047 \tag{8.54}$$

and the estimated covariance matrix of unknowns (expressed in square meters) is (Eqs. 8.42 and 8.45):

$$\begin{aligned} \hat{\Sigma}_P &= 0.0047 \cdot \begin{bmatrix} 2.20422 & 1.65766 & 2.13536 \\ 1.65766 & 1.47987 & 1.79338 \\ 2.13536 & 1.79338 & 2.31592 \end{bmatrix}^{-1} \\ &= \begin{bmatrix} 0.000092 & -0.000004 & -0.000082 \\ -0.000004 & 0.000238 & -0.000180 \\ -0.000081 & -0.000180 & 0.000224 \end{bmatrix} \end{aligned} \tag{8.55}$$

8.6 Uncertainty Evaluation When Using Adjustment Techniques

The measurement uncertainty to be associated with the unknowns estimates in Eq. 8.40 derives from the covariance matrix in Eq. 8.45.

In particular, the uncertainty of each element of $\hat{\boldsymbol{X}}$ can be obtained from the diagonal elements of matrix $\hat{\Sigma}_X \in \mathbb{R}^{N,N}$ as follows:

$$U_{x,i} = 2 \cdot \sqrt{\hat{\Sigma}_{x,i,i}} \tag{8.56}$$

A useful test for evaluating the quality of the adjustment procedure consists in comparing the a priori standard deviation (s_0) (if known) with the a posteriori one ($\tilde{s}_0$) (see Chap. 6 for an application example of such a diagnostic tool).

If the a posteriori standard deviation diverges from the a priori standard deviation, two possible sources of error are pointed out. Firstly, the stochastic model may be set up incorrectly, although it should be noted that s_0 does not affect the numerical values of the adjusted unknowns. Secondly, the functional model may be insufficient. For example, unmodelled systematic errors, or observations with gross errors, will affect the values of the unknowns.

Similarly to the case of covariance propagation, the final estimate of unknown parameter uncertainty in Eq. 8.56 is strictly related to matrix $\boldsymbol{A}$ (see Eqs. 8.42 and 8.45). This means that, if the method is applied to point localization with distributed systems, the localization uncertainty is heavily influenced by distribution within the measuring space of the network devices. In this case, the Dilution of Precision (DOP) factor is used to give a measure of the quality of the network geometry. In general, the DOP is utilized whenever approaching a localization problem based on distance measurements along lines of sight from the unknown location to reference points (Bar-Shalom et al. 2001). The DOP (or position DOP—PDOP) is defined as a function of the cofactor matrix $\boldsymbol{Q}_X$ of the unknown coordinates (see Eq. 8.42):

$$\text{DOP} = \sqrt{Tr(\boldsymbol{Q}_X)} = \sqrt{Q_{X,1,1} + Q_{X,2,2} + Q_{X,3,3}} \tag{8.57}$$

Furthermore, in addition to variances (diagonal elements of matrix $\hat{\Sigma}_X$), dependencies between adjusted parameters can also be investigated in order to assess the quality of an adjustment result. They govern the correlation coefficients in matrix $\hat{\Sigma}_X$:

$$\rho_{i,j} = \frac{\tilde{s}_{0,i,j}}{\tilde{s}_{0,i,i}\tilde{s}_{0,j,j}} \tag{8.58}$$

and hence the adequacy of the functional model and geometric configuration of the observations. Higher correlation coefficients indicate linear dependencies between parameters and they should be avoided in order to obtain robust adjustment solutions (Bar-Shalom et al. 2001).

8.7 Uncertainty Evaluation in MScMS-I Measurements

Uncertainty evaluation of 3D point coordinates measured by MScMS-I and MScMS-II can been performed using the MLPU (see Sect. 8.4). Various approaches are suggested in scientific literature according to the different problem at hand (Peggs et al. 2009). An alternative approach to MLPU, often implemented, is based on Montecarlo Sampling technique which has the potential advantage of being independent of linearization (Peggs et al. 2009).

In the present approach the MLUP has been preferred for two reasons. For LSM applications, the sensor data are typically accurate to one part in 10^4 or better, so that, despite the linearization introduced by MLUP, non-linearity in the models does not have a significant effect on the uncertainty estimates. On the other hand, uncertainty assessment in the triangulation and trilateration approaches is based on the uncertainty propagation within the least-square adjustment process of variances and covariances of input estimated parameters (see Sects. 8.5 and 8.6).

Referring to MScMS-I, the overall uncertainty of measured 3D point coordinates (i.e., the point corresponding to the probe tip) may be affected by the following contributions (see Chap. 3):

1. uncertainty of measured distances (d_i, with $i = 1, \ldots, n$) from each of the two probe Crickets (A and B) to each of the n network devices,
2. uncertainty of the localization parameters of the network devices, which is associated to the coordinates of each Cricket of the constellation,

3. uncertainty of probe geometric parameters (d(A − V) and d(A − B)),
4. Crickets synchronization error, which is considered negligible in static conditions (consideration would be necessary for a dynamic approach, i.e., in case of point tracking),
5. uncertainty of bias corrections to Cricket measurements (see Chap. 7),
6. uncertainty of 3D coordinates of probe reference points A and B ((x_A, y_A, z_A) and (x_B, y_B, z_B)), which can be traced back to the trilateration algorithm for 3D point localization.

In the following Sections the way all this contributions combine with each other is described and discussed.

8.7.1 Uncertainty of Measured Distances

The uncertainty of each measured distance of a point from the network devices may be obtained by considering the technical characteristics of Crickets (Balakrishnan et al. 2003). According to the operating manual and preliminary characterization tests, a rough estimation of standard uncertainty (i.e., standard deviation) in normal environmental conditions produces values lower than 5 mm in the whole range of measurement (see Chap. 3).

Assuming no correlation between the measured distances from a single point $\boldsymbol{x}_P \equiv (x_P, y_P, z_P)$ to the n network devices, the corresponding covariance matrix $\Sigma_d \in \mathbb{R}^{n,n}$ can be expressed as the product between a scalar σ_0^2 (variance a priori of the measured distances of the network devices) and the identity matrix $\boldsymbol{I}$:

$$\Sigma_d = \sigma_0^2 \boldsymbol{I} \tag{8.59}$$

8.7.2 Uncertainty of 3D Point Coordinates

The uncertainty of 3D point coordinates can be derived from the trilateration procedure, which consists in solving the following overdetermined system (see Sect. 3.2.2):

$$\begin{bmatrix} \sqrt{(x_1 - x_P)^2 + (y_1 - y_P)^2 + (z_1 - z_P)^2} - C_1 \\ \sqrt{(x_2 - x_P)^2 + (y_2 - y_P)^2 + (z_2 - z_P)^2} - C_2 \\ \vdots \\ \sqrt{(x_n - x_P)^2 + (y_n - y_P)^2 + (z_n - z_P)^2} - C_n \end{bmatrix} = \begin{bmatrix} d_1 \\ d_2 \\ \vdots \\ d_n \end{bmatrix} \tag{8.60}$$

where (x_P, y_P, z_P) are the unknown coordinates of point $\boldsymbol{x}_P \equiv (x_P, y_P, z_P)$ to be localized, $n > 3$ is the total number of network devices in connection with the Cricket positioned in $\boldsymbol{x}_P \equiv (x_P, y_P, z_P)$, (x_1, y_1, z_1), (x_2, y_2, z_2), …, (x_n, y_n, z_n) are the n set of coordinates of the network devices, known by system calibration (see Chap. 5), d_1, d_2, …, d_n are the measured distances from point $\boldsymbol{x}_P \equiv (x_P, y_P, z_P)$ to the network devices, C_1, C_2, …, C_n are the bias corrections of distance measurements, obtained by implementing a performance enhancing procedure (see Chap. 7).

The first step of the described approach consist in the linearization of the Eq. 8.60. Afterwards the system is solved by least-squares adjustment (in this way, known variable may be weighted in least-squares adjustment according to the related covariance matrix), and then the calculation of the uncertainty of coordinates is automatically performed.

For each point to be localized, the linearized expression of the trilateration equations is (see Eq. 8.31):

$$\boldsymbol{A}\Delta\hat{\boldsymbol{x}}_P = \Delta\boldsymbol{d} + \boldsymbol{v} \tag{8.61}$$

where $\Delta\boldsymbol{d}$ is the vector of reduced observations (see Eq. 8.28), obtained by calculating the left term of Eq. 8.60 for the approximations (x_P^0, y_P^0, z_P^0) for 3D coordinates of the point to be localized (as well as for the known values of the $3 \times n$ coordinates of the network devices and the bias corrections of distance measurements), and subtracting it to the observations (i.e., the measured distances d_1, d_2, …, d_n):

$$\begin{bmatrix} d_1 - \sqrt{(x_1 - x_P^0)^2 + (y_1 - y_P^0)^2 + (z_1 - z_P^0)^2} + C_1 \\ d_2 - \sqrt{(x_2 - x_P^0)^2 + (y_2 - y_P^0)^2 + (z_2 - z_P^0)^2} + C_2 \\ \vdots \\ d_n - \sqrt{(x_n - x_P^0)^2 + (y_n - y_P^0)^2 + (z_n - z_P^0)^2} + C_n \end{bmatrix} = \begin{bmatrix} \Delta d_1 \\ \Delta d_2 \\ \vdots \\ \Delta d_n \end{bmatrix} \tag{8.62}$$

$\boldsymbol{v}$ is the vector of the residuals of distances between the point to be localized and the network devices, $\Delta\hat{\boldsymbol{x}}_P$ is the vector of the estimates of the reduced unknowns, corresponding to the difference between the estimated values of the 3D coordinates of the point to be localized and their approximations $\boldsymbol{x}_P^0 = \begin{bmatrix} x_P^0 & y_P^0 & z_P^0 \end{bmatrix}^T$:

$$\Delta\hat{\boldsymbol{x}}_P = \begin{bmatrix} \hat{x}_P - x_P^0 \\ \hat{y}_P - y_P^0 \\ \hat{z}_P - z_P^0 \end{bmatrix} \tag{8.63}$$

$\boldsymbol{A} \in \mathbb{R}^{n,3}$ is the matrix of partial derivatives (Jacobian) of the functions in the left side of Eq. 8.60 with respect to each of the 3D point coordinates (calculated in $\boldsymbol{x}_P^0 = \begin{bmatrix} x_P^0 & y_P^0 & z_P^0 \end{bmatrix}^T$),

$$
\boldsymbol{A} = \begin{bmatrix} \frac{-(x_1 - x_P^0)}{G_1^0} & \frac{-(y_1 - y_P^0)}{G_1^0} & \frac{-(z_1 - z_P^0)}{G_1^0} \\ \frac{-(x_2 - x_P^0)}{G_2^0} & \frac{-(y_2 - y_P^0)}{G_2^0} & \frac{-(z_2 - z_P^0)}{G_2^0} \\ \vdots & \vdots & \vdots \\ \frac{-(x_n - x_P^0)}{G_n^0} & \frac{-(y_n - y_P^0)}{G_n^0} & \frac{-(z_n - z_P^0)}{G_n^0} \end{bmatrix} \tag{8.64}
$$

where:

$$
\begin{bmatrix} \sqrt{(x_1 - x_P^0)^2 + (y_1 - y_P^0)^2 + (z_1 - z_P^0)^2} \\ \sqrt{(x_2 - x_P^0)^2 + (y_2 - y_P^0)^2 + (z_2 - z_P^0)^2} \\ \vdots \\ \sqrt{(x_n - x_P^0)^2 + (y_n - y_P^0)^2 + (z_n - z_P^0)^2} \end{bmatrix} = \begin{bmatrix} G_1^0 \\ G_2^0 \\ \vdots \\ G_n^0 \end{bmatrix} \tag{8.65}
$$

The resulting normal equation (see Eq. 8.37) is:

$$
\boldsymbol{N}\Delta\hat{\boldsymbol{x}}_P = \boldsymbol{T} \tag{8.66}
$$

where:

$$
\boldsymbol{N} = \boldsymbol{A}^T \boldsymbol{P_d} \boldsymbol{A} \tag{8.67}
$$

and

$$
\boldsymbol{T} = \boldsymbol{A}^T \boldsymbol{P_d} \Delta \boldsymbol{D} \tag{8.68}
$$

$\boldsymbol{P_d} \in \mathbb{R}^{n,n}$ is the weight matrix, obtained as the inverse of the covariance matrix $\Sigma_{\boldsymbol{d}} \in \mathbb{R}^{n,n}$ associated with the measured distances. If the elements of this matrix are unknowns it is assumed equal to the identity matrix $\boldsymbol{I}$. That means that all the measured distances are assumed to have the same weight. This assumption is legitimate when there is no explicit reason to assign different weights.

The general solution of Eq. 8.66 in order to estimate the unknown vector $\Delta\hat{\boldsymbol{x}}_P$ is:

$$
\Delta\hat{\boldsymbol{x}}_P = \boldsymbol{N}^{-1}\boldsymbol{T} \tag{8.69}
$$

If the covariance matrices for distance measurements $(\Sigma_{\boldsymbol{d}} \in \mathbb{R}^{n,n})$, coordinates of network devices $(\Sigma_{\boldsymbol{B}} \in \mathbb{R}^{3n,3n})$ and bias corrections $(\Sigma_{\boldsymbol{C}} \in \mathbb{R}^{\ltimes,\ltimes})$ are known, the overall covariance $\Sigma_x \in \mathbb{R}^{3,3}$ of the 3D point coordinates can be estimated by applying the MLPU to the following linearized expression of Eq. 8.60:

$$
\boldsymbol{J_B}\Delta\boldsymbol{x_B} + \boldsymbol{J_C}\Delta\boldsymbol{C} + \boldsymbol{A}\Delta\hat{\boldsymbol{x}}_P = \Delta\boldsymbol{d} + \boldsymbol{v} \tag{8.70}
$$

where $\boldsymbol{J_B} \in \mathbb{R}^{n,3n}$ is the matrix of partial derivatives (Jacobian) of the n functions in the left side of Eq. 8.60 with respect to each of the 3 coordinates of the n network devices (calculated for the approximated values of the coordinates of the point to be localized and for the known values of the coordinates of the network devices and the bias corrections),

$$\boldsymbol{J_B} = \begin{bmatrix} \frac{(x_1 - x_P^0)}{G_1^0} & \frac{(y_1 - y_P^0)}{G_1^0} & \frac{(z_1 - z_P^0)}{G_1^0} & 0 & 0 & 0 & 0 \\ \vdots & \vdots & \vdots & \ddots & \vdots & \vdots & \vdots \\ 0 & 0 & 0 & 0 & \frac{(x_n - x_P^0)}{G_n^0} & \frac{(y_n - y_P^0)}{G_n^0} & \frac{(z_n - z_P^0)}{G_n^0} \end{bmatrix} \tag{8.71}$$

$\boldsymbol{J_C} \in \mathbb{R}^{n,n}$ is the matrix of partial derivatives (Jacobian) of the n functions in the left side of Eq. 8.60 with respect to each bias correction of the n distance measurements (calculated for the approximated values of the coordinates of the point to be localized and for the known values of the coordinates of the network devices and the bias corrections),

$$\boldsymbol{J_C} = \begin{bmatrix} -1 & 0 & \cdots & 0 \\ 0 & -1 & \cdots & 0 \\ \vdots & \vdots & \ddots & 0 \\ 0 & 0 & 0 & -1 \end{bmatrix} \tag{8.72}$$

$\boldsymbol{\Delta x_B}$ and $\boldsymbol{\Delta C}$ are respectively the vectors of the variations of the 3D coordinates of the n network devices and the n bias corrections from their actual values (in this approximation both vectors have all the elements equal to 0).

In general, an estimate of matrix $\boldsymbol{\Sigma_B}$ is obtained by the calibration procedure (see Chap. 5). It is a block 3×3 diagonal matrix, with all the elements out of the block diagonal equal to 0. That means that the coordinates of a given network device are correlated to each others, but they are not correlated to the coordinates of the other network devices. This assertion holds if the positions of each network device have been calibrated separately from that of other network devices. If global calibration algorithms (i.e., algorithms optimizing the positions of the network devices of the whole system) have been implemented, the statement does not hold any more and there is no guarantee of no correlation between coordinates of different network devices.

However, in practical applications, it is usually assumed that $\boldsymbol{\Sigma_B}$ is a $3n \times 3n$ diagonal matrix, with all the elements out of the diagonal equal to 0.

In the same way $\boldsymbol{\Sigma_d}$ and $\boldsymbol{\Sigma_C}$ are $n \times n$ diagonal matrices, with all the elements out of the diagonal equal to 0. The elements of $\hat{\boldsymbol{\Sigma}}_{\boldsymbol{d}}$ are obtained from Crickets manual or applying specific calibration procedure (see Sect. 8.7.1). Matrix $\hat{\boldsymbol{\Sigma}}_{\boldsymbol{C}}$ is derived from the procedure for the calculation of bias corrections (see Chap. 7).

Table 8.3 Example of 6 distances measured from a Cricket (see Chap. 3) positioned inside a measurement space and 6 network devices positioned all around it

	Meas. distance d_i (m)	Correction C_i (m)	x_i (m)	y_i (m)	z_i (m)
Device 1	2.454	0.004	1.500	1.500	2.500
Device 2	3.010	0.004	1.500	2.500	2.500
Device 3	3.005	0.005	2.500	1.500	2.500
Device 4	3.474	0.006	2.500	2.500	2.500
Device 5	3.749	0.005	3.500	1.500	2.500
Device 6	4.127	0.004	3.500	2.500	2.500

The least-squares adjustment applied to Eq. 8.70 gives the following solution:

$$\Delta\hat{\boldsymbol{x}}_P = \boldsymbol{N}^{-1}\left(\boldsymbol{T} - \boldsymbol{A}^T\hat{\boldsymbol{P}}_d\boldsymbol{J}_B\Delta\boldsymbol{x}_B - \boldsymbol{A}^T\hat{\boldsymbol{P}}_d\boldsymbol{J}_C\Delta\boldsymbol{C}\right) \tag{8.73}$$

Hence, the combined point estimated covariance (calculated by applying the MLPU to Eq. 8.73) is:

$$\begin{aligned}\hat{\Sigma}_P = \boldsymbol{N}^{-1} + \boldsymbol{N}^{-1}\left(\boldsymbol{A}^T\hat{\boldsymbol{P}}_d\boldsymbol{J}_B\right)\hat{\Sigma}_B\left(\boldsymbol{A}^T\hat{\boldsymbol{P}}_d\boldsymbol{J}_B\right)^T\boldsymbol{N}^{-1} \\ + \boldsymbol{N}^{-1}\left(\boldsymbol{A}^T\hat{\boldsymbol{P}}_d\boldsymbol{J}_C\right)\hat{\Sigma}_C\left(\boldsymbol{A}^T\hat{\boldsymbol{P}}_d\boldsymbol{J}_C\right)^T\boldsymbol{N}^{-1}\end{aligned} \tag{8.74}$$

In order to better understand the procedure, consider again the example reported at the end of Sect. 8.5. $n = 6$ distances (reported in Table 8.3) have been measured from a Cricket positioned within a measuring volume and 6 network devices positioned all around it. The 3D coordinates of each network device are known, as well as the bias corrections to the measured distances (Table 8.3).

The estimated covariance matrices of the measured distances, the bias corrections of distance measurements, and the coordinates of the network devices are, respectively (values expressed in m^2):

$$\hat{\Sigma}_d = 10^{-7}\cdot\begin{bmatrix}16 & 0 & 0 & 0 & 0 & 0\\ 0 & 13 & 0 & 0 & 0 & 0\\ 0 & 0 & 24 & 0 & 0 & 0\\ 0 & 0 & 0 & 20 & 0 & 0\\ 0 & 0 & 0 & 0 & 18 & 0\\ 0 & 0 & 0 & 0 & 0 & 25\end{bmatrix} \tag{8.75}$$

$$\hat{\Sigma}_C = 10^{-7}\cdot\begin{bmatrix}16 & 0 & 0 & 0 & 0 & 0\\ 0 & 18 & 0 & 0 & 0 & 0\\ 0 & 0 & 13 & 0 & 0 & 0\\ 0 & 0 & 0 & 15 & 0 & 0\\ 0 & 0 & 0 & 0 & 18 & 0\\ 0 & 0 & 0 & 0 & 0 & 11\end{bmatrix} \tag{8.76}$$

$$\hat{\Sigma}_B = 10^{-8} \cdot \begin{bmatrix} 5 & 0 & 0 & 0 & 0 & 0 & 0 & 0 & 0 & 0 & 0 & 0 & 0 & 0 & 0 & 0 & 0 & 0 \\ 0 & 5 & 0 & 0 & 0 & 0 & 0 & 0 & 0 & 0 & 0 & 0 & 0 & 0 & 0 & 0 & 0 & 0 \\ 0 & 0 & 5 & 0 & 0 & 0 & 0 & 0 & 0 & 0 & 0 & 0 & 0 & 0 & 0 & 0 & 0 & 0 \\ 0 & 0 & 0 & 2 & 0 & 0 & 0 & 0 & 0 & 0 & 0 & 0 & 0 & 0 & 0 & 0 & 0 & 0 \\ 0 & 0 & 0 & 0 & 2 & 0 & 0 & 0 & 0 & 0 & 0 & 0 & 0 & 0 & 0 & 0 & 0 & 0 \\ 0 & 0 & 0 & 0 & 0 & 2 & 0 & 0 & 0 & 0 & 0 & 0 & 0 & 0 & 0 & 0 & 0 & 0 \\ 0 & 0 & 0 & 0 & 0 & 0 & 5 & 0 & 0 & 0 & 0 & 0 & 0 & 0 & 0 & 0 & 0 & 0 \\ 0 & 0 & 0 & 0 & 0 & 0 & 0 & 5 & 0 & 0 & 0 & 0 & 0 & 0 & 0 & 0 & 0 & 0 \\ 0 & 0 & 0 & 0 & 0 & 0 & 0 & 0 & 5 & 0 & 0 & 0 & 0 & 0 & 0 & 0 & 0 & 0 \\ 0 & 0 & 0 & 0 & 0 & 0 & 0 & 0 & 0 & 8 & 0 & 0 & 0 & 0 & 0 & 0 & 0 & 0 \\ 0 & 0 & 0 & 0 & 0 & 0 & 0 & 0 & 0 & 0 & 8 & 0 & 0 & 0 & 0 & 0 & 0 & 0 \\ 0 & 0 & 0 & 0 & 0 & 0 & 0 & 0 & 0 & 0 & 0 & 8 & 0 & 0 & 0 & 0 & 0 & 0 \\ 0 & 0 & 0 & 0 & 0 & 0 & 0 & 0 & 0 & 0 & 0 & 0 & 2 & 0 & 0 & 0 & 0 & 0 \\ 0 & 0 & 0 & 0 & 0 & 0 & 0 & 0 & 0 & 0 & 0 & 0 & 0 & 2 & 0 & 0 & 0 & 0 \\ 0 & 0 & 0 & 0 & 0 & 0 & 0 & 0 & 0 & 0 & 0 & 0 & 0 & 0 & 2 & 0 & 0 & 0 \\ 0 & 0 & 0 & 0 & 0 & 0 & 0 & 0 & 0 & 0 & 0 & 0 & 0 & 0 & 0 & 7 & 0 & 0 \\ 0 & 0 & 0 & 0 & 0 & 0 & 0 & 0 & 0 & 0 & 0 & 0 & 0 & 0 & 0 & 0 & 7 & 0 \\ 0 & 0 & 0 & 0 & 0 & 0 & 0 & 0 & 0 & 0 & 0 & 0 & 0 & 0 & 0 & 0 & 0 & 7 \end{bmatrix} \tag{8.77}$$

The objective is to determine the 3D coordinates of the Cricket (x_P, y_P, z_P) and the corresponding covariance matrix $\hat{\Sigma}_P$.

Fixing at 0 the approximations of 3D coordinates of the point to be localized, we obtain the following vector of unknowns:

$$\Delta \boldsymbol{x}_P = \begin{bmatrix} \Delta x_P & \Delta y_P & \Delta z_P \end{bmatrix}^T = \begin{bmatrix} x_P & y_P & z_P \end{bmatrix}^T \tag{8.78}$$

The corresponding design matrix $\boldsymbol{A}$ is:

$$\boldsymbol{A} = \begin{bmatrix} -0.45750 & -0.45750 & -0.76249 \\ -0.39057 & -0.65094 & -0.65094 \\ -0.65094 & -0.39057 & -0.65094 \\ -0.57735 & -0.57735 & -0.57735 \\ -0.76835 & -0.32929 & -0.54882 \\ -0.70353 & -0.50252 & -0.50252 \end{bmatrix} \tag{8.79}$$

The matrix of normal equations $\boldsymbol{N}$ and the absolute term vector $\boldsymbol{T}$ are respectively:

$$\boldsymbol{N} = \begin{bmatrix} 111733.29583 & 88095.63587 & 113249.52959 \\ 88095.63587 & 84823.62483 & 101797.90745 \\ 113249.52959 & 101797.90745 & 130088.37849 \end{bmatrix} \tag{8.80}$$

$$\boldsymbol{T} = \begin{bmatrix} 154188.3339 \\ 135275.7790 \\ 169894.6794 \end{bmatrix} \tag{8.81}$$

The Jacobians $\boldsymbol{J_B}$ and $\boldsymbol{J_C}$ are respectively:

$$\boldsymbol{J_B} = \begin{bmatrix} 0.45750 & 0 & 0 & 0 & 0 & 0 \\ 0.45750 & 0 & 0 & 0 & 0 & 0 \\ 0.76249 & 0 & 0 & 0 & 0 & 0 \\ 0 & 0.39057 & 0 & 0 & 0 & 0 \\ 0 & 0.65094 & 0 & 0 & 0 & 0 \\ 0 & 0.65094 & 0 & 0 & 0 & 0 \\ 0 & 0 & 0.65094 & 0 & 0 & 0 \\ 0 & 0 & 0.39057 & 0 & 0 & 0 \\ 0 & 0 & 0.65094 & 0 & 0 & 0 \\ 0 & 0 & 0 & 0.57735 & 0 & 0 \\ 0 & 0 & 0 & 0.57735 & 0 & 0 \\ 0 & 0 & 0 & 0.57735 & 0 & 0 \\ 0 & 0 & 0 & 0 & 0.76834 & 0 \\ 0 & 0 & 0 & 0 & 0.32929 & 0 \\ 0 & 0 & 0 & 0 & 0.54882 & 0 \\ 0 & 0 & 0 & 0 & 0 & 0.70352 \\ 0 & 0 & 0 & 0 & 0 & 0.50252 \\ 0 & 0 & 0 & 0 & 0 & 0.50252 \end{bmatrix}^T \quad (8.82)$$

$$\boldsymbol{J_C} = \begin{bmatrix} -1 & 0 & 0 & 0 & 0 & 0 \\ 0 & -1 & 0 & 0 & 0 & 0 \\ 0 & 0 & -1 & 0 & 0 & 0 \\ 0 & 0 & 0 & -1 & 0 & 0 \\ 0 & 0 & 0 & 0 & -1 & 0 \\ 0 & 0 & 0 & 0 & 0 & -1 \end{bmatrix} \quad (8.83)$$

Hence, by applying Eqs. 8.48 and 8.53, we obtain the final results:

$$\hat{\boldsymbol{x}}_P = \Delta\hat{\boldsymbol{x}}_P = [0.4983 \quad 0.5016 \quad 0.4797]^T \quad (8.84)$$

$$\hat{\Sigma}_P = \begin{bmatrix} 0.00001478 & -0.00000037 & -0.00001303 \\ -0.00000037 & 0.00003844 & -0.00002950 \\ -0.00001303 & -0.00002950 & 0.00003615 \end{bmatrix} \quad (8.85)$$

8.7.3 Uncertainty of Probe Tip Coordinates

Under the hypothesis that the measurements of the coordinates of the two probe Crickets and the two geometrical parameters of the probe (d(A − V) and d(A − B)) are independent, the covariance of the probe tip coordinates $\Sigma_V \in \mathbb{R}^{3,3}$ in each measurement is estimated by applying the MLPU to the equation for estimating the coordinates of the probe tip (see Chap. 3):

$$\hat{\Sigma}_V = \boldsymbol{J}\Sigma_E\boldsymbol{J}^T \tag{8.86}$$

where $\boldsymbol{J} \in \mathbb{R}^{3,8}$ is the Jacobian of the functional for estimating the coordinates of the probe tip (see Eq. 3.6), i.e., the matrix of its partial derivatives with respect to the 3 coordinates of the centres of the two probe Crickets, and the parameters $d(\mathrm{A}-\mathrm{V})$ and $d(\mathrm{A}-\mathrm{B})$:

$$\boldsymbol{J} = \begin{bmatrix} 1-\frac{d(A-V)}{d(A-B)} & 0 & 0 \\ 0 & 1-\frac{d(A-V)}{d(A-B)} & 0 \\ 0 & 0 & 1-\frac{d(A-V)}{d(A-B)} \\ \frac{d(A-V)}{d(A-B)} & 0 & 0 \\ 0 & \frac{d(A-V)}{d(A-B)} & 0 \\ 0 & 0 & \frac{d(A-V)}{d(A-B)} \\ \frac{(x_B-x_A)}{d(A-B)} & \frac{(y_B-y_A)}{d(A-B)} & \frac{(z_B-z_A)}{d(A-B)} \\ \frac{(x_B-x_A)d(A-V)}{(d(A-B))^2} & \frac{(y_B-y_A)d(A-V)}{(d(A-B))^2} & \frac{(z_B-z_A)d(A-V)}{(d(A-B))^2} \end{bmatrix}^T \tag{8.87}$$

$\Sigma_E \in \mathbb{R}^{8,8}$ is the covariance matrix of parameters in the equation for estimating the coordinates of the probe tip, obtained by composing in a unique covariance matrix the covariance matrices of the coordinates of point A (Σ_A) and point B (Σ_B) (obtained by applying the procedure explained in Sect. 8.7.2), and adding the variances $\sigma^2_{d(A-V)}$ and $\sigma^2_{d(A-B)}$ corresponding to $d(A-V)$ and $d(A-B)$ respectively (estimated during the probe calibration):

$$\Sigma_E = \begin{bmatrix} \Sigma_\mathrm{A} & \mathbf{0} & \mathbf{0} & \mathbf{0} \\ \mathbf{0} & \Sigma_\mathrm{B} & \mathbf{0} & \mathbf{0} \\ \mathbf{0} & \mathbf{0} & \sigma^2_{d(A-V)} & \mathbf{0} \\ \mathbf{0} & \mathbf{0} & \mathbf{0} & \sigma^2_{d(A-B)} \end{bmatrix} \tag{8.88}$$

The diagonal elements of $\hat{\Sigma}_V$ are the estimates of the variances of the 3D coordinates of the point measured by the probe. They can be used to evaluate the expanded uncertainty of point coordinates by extracting the corresponding standard deviations and multiplying them by an opportune coverage factor (usually equal to 2) (JCGM 100:2008 2008):

$$\begin{aligned} U_{\mathrm{V},x} &= 2\cdot\sqrt{\hat{\Sigma}_{\mathrm{V},1,1}} \\ U_{\mathrm{V},y} &= 2\cdot\sqrt{\hat{\Sigma}_{\mathrm{V},2,2}} \\ U_{\mathrm{V},z} &= 2\cdot\sqrt{\hat{\Sigma}_{\mathrm{V},3,3}} \end{aligned} \tag{8.89}$$

The corresponding 3D radial uncertainty is:

$$U_V = 2 \cdot \sqrt{\hat{\Sigma}_{V,1,1} + \hat{\Sigma}_{V,2,2} + \hat{\Sigma}_{V,3,3}} \tag{8.90}$$

The uncertainty contribution due to the correction of the non-punctiform shape of the tip is usually neglected. Therefore, it can be considered by adding minor modifications to Eq. 8.88.

Consider for example the MScMS-I prototype, characterized by $d(A-V) = 540$ mm and $d(A-B) = 450$ mm, and the corresponding estimates of standard uncertainties, respectively $\hat{\sigma}_{d(A-V)} = 0.01$ mm and $\hat{\sigma}_{d(A-B)} = 0.01$ mm.

Measuring a given point in the space, the following two covariance matrices of Cricket A ($\boldsymbol{x}_A = (1.500, 1.300, 2.000)$) and B ($\boldsymbol{x}_B = (1.700, 0.896, 2.000)$) are estimated (values expressed in m^2):

$$\hat{\Sigma}_A = \begin{bmatrix} 0.000015 & -0.000001 & -0.000012 \\ -0.000001 & 0.000035 & -0.000030 \\ -0.000012 & -0.000030 & 0.000035 \end{bmatrix} \tag{8.91}$$

$$\hat{\Sigma}_B = \begin{bmatrix} 0.000020 & -0.000005 & -0.000020 \\ -0.000005 & 0.000020 & -0.000015 \\ -0.000020 & -0.000015 & 0.000025 \end{bmatrix} \tag{8.92}$$

The Jacobian of the functional for estimating the coordinates of the probe tip (see Eq. 8.87) results to be:

$$\boldsymbol{J} = \begin{bmatrix} -0.1979 & 0 & 0 & 1.1979 & 0 & 0 & 0.4437 & 0.5315 \\ 0 & -0.1979 & 0 & 0 & 1.1979 & 0 & -0.8962 & \\ 0 & 0 & -0.1979 & 0 & 0 & 1.1979 & 0 & 0 \end{bmatrix} \tag{8.93}$$

The estimate of the parameters covariance matrix becomes:

$$\hat{\Sigma}_E = 10^{-6} \cdot \begin{bmatrix} 15 & -1 & -12 & 0 & 0 & 0 & 0 & 0 \\ -1 & 35 & -30 & 0 & 0 & 0 & 0 & 0 \\ -12 & -30 & 35 & 0 & 0 & 0 & 0 & 0 \\ 0 & 0 & 0 & 20 & -5 & -20 & 0 & 0 \\ 0 & 0 & 0 & -5 & 20 & -15 & 0 & 0 \\ 0 & 0 & 0 & -20 & -15 & 25 & 0 & 0 \\ 0 & 0 & 0 & 0 & 0 & 0 & 0.0001 & 0 \\ 0 & 0 & 0 & 0 & 0 & 0 & 0 & 0.0001 \end{bmatrix} \tag{8.94}$$

Hence, after applying Eq. 8.86, the estimated covariance matrix of the probe tip coordinates is:

$$\hat{\Sigma}_V = \begin{bmatrix} 0.0000292859 & -0.0000072138 & -0.0000291684 \\ -0.0000072138 & 0.0000300693 & -0.0000226986 \\ -0.0000291684 & -0.0000226986 & 0.0000372437 \end{bmatrix} \tag{8.95}$$

This produces the following expanded uncertainty of point coordinates and 3D radial uncertainty (values expressed in m):

$$\begin{aligned} U_{V,x} &= 0.011 \\ U_{V,y} &= 0.011 \\ U_{V,z} &= 0.012 \end{aligned} \quad (8.96)$$

$$U_V = 0.020 \quad (8.97)$$

In this example, the achieved result shows that the expected values of the overall uncertainty are in the range of 1 cm. This high value is mainly caused by the distance measurements obtained by US transducers. They give the major contribution to the combined uncertainty by the MLPU approach (see Eq. 8.94). On the other hand, the contributions coming from the uncertainties of probe geometrical parameters are minor in comparison to the others (see Eq. 8.94).

Even if this overall uncertainty value may be acceptable for many applications in LSM, for practical applications a substantial reduction is required, for example, by improving the performance of the network devices, or replacing the US sensors with a more accurate technology.

8.8 Uncertainty Evaluation in MScMS-II Measurements

As mentioned in Sect. 8.7, the uncertainty evaluation of measured 3D point coordinates by MScMS-II is performed using the MLPU (JCGM 100:2008 2008).

According to the system architecture (see Chap. 3), the main contributions to overall uncertainty of 3D point coordinates (i.e., probe tip coordinates) may be traced in the following sources:

1. uncertainty of 2D point coordinates, which refers to the 2D pixel coordinates of point projection in the camera view plane,
2. uncertainty of camera calibration parameters, which is associated to the internal and external camera parameters obtained in the calibration phase,
3. camera synchronization error, which is considered negligible in static conditions (consideration would be necessary for a dynamic approach, i.e., in case of point tracking),
4. uncertainty of 3D point coordinates, which can be traced back to the triangulation algorithm for 3D point reconstruction.

8.8.1 Uncertainty of 2D Point Coordinates

Assuming that the uncertainty in each 2D point coordinates is uncorrelated with that of any other point coordinates (this is true in absence of systematic errors), both other points imaged on the same camera image and other camera images of the same point, only uncertainty in the coordinates of any single image point may be correlated (Mikhail et al. 2001). Hence, each couple of coordinates $(u_{i,j}, v_{i,j})$ in the 2D image plane of a given camera i (see Sect. 3.3.3), corresponding to a specific point j in the 3D space, can be associated to a covariance matrix $\Sigma_{i,j} \in \mathbb{R}^{2,2}$.

According to the MScMS-II working principles, the centres of two spherical markers must be localized (see Chap. 3). In this case, the main factors that may contribute to form $\Sigma_{i,j}$ matrix are: camera technical characteristics (resolution, focal length, FOV, sensitivity, lens distortion), system layout geometry (size of markers, distance between camera and markers, position of markers with respect to the normal of the camera plane of view, fraction of marker surface visible by each camera), computational procedures (2D point correspondence analysis, image processing algorithms), and noise or false-measurements (e.g., camera vibrations, external IR sources, reflections).

All these factors give their direct or indirect contribution to the uncertainty in the measurement of 2D point coordinates. A rough estimate of 2D point uncertainty may be obtained through the pixel accuracy of CCD cameras. In general, considering the image plane of a camera, the average measurement accuracy of non-signalized points is around 0.2–0.5 pixel (Luhmann et al. 2006). If the feature to be measured consists of a symmetrical distribution of pixel values, many mathematical methods can be implemented to determine the centre. Among these worth mentioning: the Local Centroid method, the Correlation method, the Least-Squares Matching method, and the Structural methods (Luhmann et al. 2006). All of them may be efficiently implemented in order to obtain centre point coordinates and the corresponding uncertainty. The Least-Squares Matching method, however, is the most convenient since it automatically provides the covariance matrix of the two calculated coordinates.

A significant factor in determining 2D point uncertainty is the size (diameter) of imaged points. It is directly correlated to marker position and distance, and camera characteristics (resolution, focal length, Field Of View, sensitivity and lens distortion). The optimum marker size ranges between about 5 and 15 pixels in diameter (Luhmann et al. 2006). Smaller points do not provide enough information, while larger points diameters may result in too large numbers of observations to be processed.

An empirical relationship between marker size and achievable point accuracy, reported by Luhman et al. (see Fig. 8.2), shows that as the marker size increases over 25 pixels, the function rapidly converges to an accuracy of about 0.005 pixel (Luhmann et al. 2006). On the other hand, with a marker size under 5 pixels, the accuracy reaches values over 0.01 pixel.

For a preliminary assessment of point coordinate uncertainty in MScMS-II measurements, considering the low-level resolution of the cameras used for the prototype assembling, the pixel uncertainty (intended as standard deviation of each of the two pixel coordinates) has been assumed equal to 0.5 pixel with no covariance between the two coordinates.

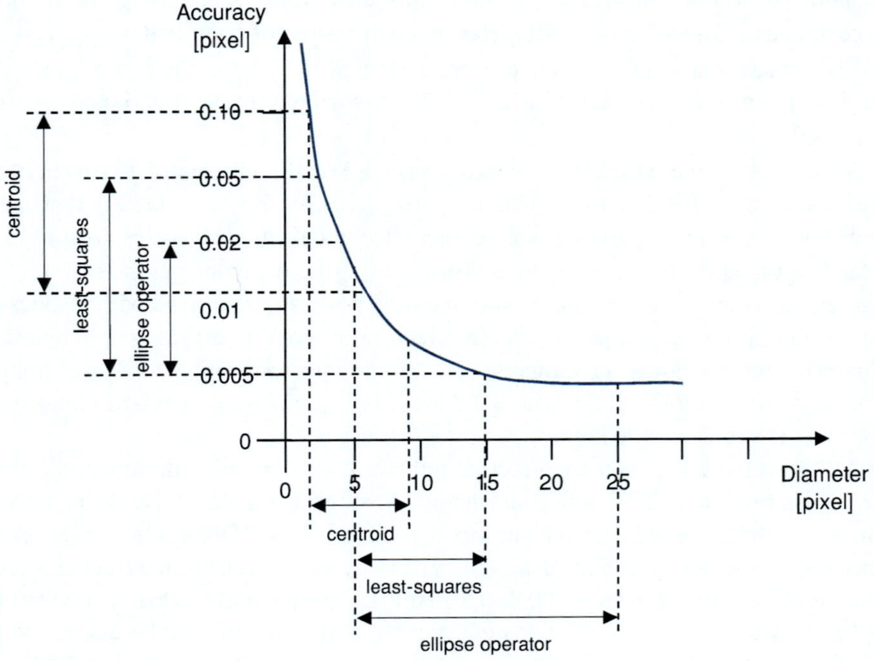

Fig. 8.2 Practical use and accuracy potential of different point measurement operators as a function of marker diameter (*y axis* not to scale) (adapted from Luhmann et al. 2006). (with permission)

In standard operational condition of the system, a Matlab® routine for image processing, based on Least-Squares Matching method, provides the covariance matrices for the centres of each of the two probe markers, as well as the 2D coordinates.

8.8.2 Uncertainty of 3D Point Coordinates

The uncertainty of 3D point coordinates can be derived by applying the MLPU (JCGM 100:2008 2008) to the Collinearity Equations, which are the basis of photogrammetry theory and originate from the perspective projection of a point in the 3D space onto a given camera viewing plane (see Sect. 3.3.3).

This approach produces the covariance matrix for each point localized by the system.

For a complete description of this technique, which is similar to that presented in Sect. 8.7, readers are invited to review specialist literature (Mikhail et al. 2001; Luhmann et al. 2006).

8.8.3 Uncertainty of Probe Tip Coordinates

Similarly to the case of MScMS-I, the covariance of the probe tip coordinates $\Sigma_V \in \mathbb{R}^{3,3}$ in each measurement is estimated by applying the MLPU to the linearized form of the equation for estimating the coordinates of the probe tip (see Eq. 3.6). The procedure is totally equivalent to that exposed in Sect. 8.7.3 referring to MScMS-I.

References

3rdTech Inc. (2010) http://www.3rdtech.com/. Accessed date: 15/10/2010

Adler D (1993) Genetic algorithms and simulated annealing: a marriage proposal. IEEE International Conference on Neural Networks San Francisco, CA, 28 March–1 April

Ai J, Abouzeid AA (2006) Coverage by directional sensors in randomly deployed wireless sensor networks. J Combin Optimiz 11:21–41

Akcan H, Kriakov V, Brönnimann H, Delis A (2006) GPS-free node localization in mobile wireless sensor networks. In: Proceedings of MobiDE'06, Chicago, IL, USA, pp 35–42

Akyildiz IF, Su W, Sankarasubramaniam Y, Cayirci E (2002) A survey on sensor networks. IEEE Commun Mag 40:102–114

Alici G, Shirinzadeh B (2003) Laser interferometry based robot position error modelling for kinematic calibration. In: Proceedings of the 2003 IEEE/RSJ, international conference on intelligent robots and systems, Las Vegas, Nevada Oct 2003

Andersen R. (2008) Modern methods for robust regression. Sage University Paper Series on Quantitative Applications in the Social Sciences, 07-152

ANSI/ASME B89.4.19-2006 (2006) Performance evaluation of laser-based spherical coordinate measurement systems, New York

ARC Second (2010). Product literature, http://arcsecond.com. Accessed date: 15/10/2010

Automated Precision Inc. (2010) http://www.apisensor.com/tracker3.html. Accessed date: 15/10/2010

Axios 3D Services GmbH (2010) http://www.axios3d.de/. Accessed date: 15/10/2010

Balakrishnan H, Baliga R, Curtis D, Goraczko M, Miu A, Priyantha N B, Smith A, Steele K, Teller S, Wang K (2003) Lessons from developing and deploying the cricket indoor location system. MIT Tech Rep

Balicki J, Kitowski Z (2001) Multicriteria evolutionary algorithm with Tabu search for task assignment. Lecture notes in computer science, (1993/2001) pp 373–384

Bar-Shalom Y, Li XR, Kirubarajan T (2001) Estimation with applications to tracking and navigation. Wiley, New York

Berners AC, Webster JG, Worringham CJ, Stelmach GE (1995) An ultrasonic time-of-flight system for hand movement measurement. Physiolo Meas 16:203–211

Berry DW (1961) Precise surveying in the construction of Nimrod. The Charterer Surveyor, April 1961

Biagioni E S, Sasaki G (2003) Wireless sensor placement for reliable and efficient data collection. In: Proceedings of IEEE Hawaii international conference on system sciences (HICSS), Hawaii, USA, Jan 2003

Blender Foundation (2008). http://www.blender.org. Accessed 25 October 2008

Bogdanov A, Maneva E, Riesenfeld S (2004) Power-aware base station positioning for sensor networks. In: Proceedings of 23rd international annual joint conference of the IEEE computer and communications societies (INFOCOM'04), Hong Kong, Mar 2004

Bohn DA (1988) Environmental effects on the speed of sound. J Audio Eng Soc 36(4):223–231

Bosch JA (1995) Coordinate measuring machines and systems. Ed. Marcel Dekker, New York

Box GEP, Hunter WG, Hunter JS (1978) Statistics for experimenters. Wiley, New York

Brown DC (1971) Close-range camera calibration. Photogramm Eng 37:855–866

Cassandras CG, Li W (2005) Sensor networks and cooperative control. Eur J Control 11:436–463

Cauchick-Miguel P, King T, Davis J (1996) CMM verification: a survey. Measurement 17(1):1–16

Chakrabarty K, Iyengar S, Qi H, Cho E (2002) Grid coverage or surveillance and target location in distributed sensor networks. IEEE Trans Mob Comput 51:1448–1453

Chen M, Cheng F, Gudavalli R (2003) Precision and accuracy in an lndoor localization system. Technical report CS294–1/2, University of California, Berkeley, USA

Chen Y, Chuan C, Zhao Q (2005) Sensor placement for maximizing lifetime per unit cost in wireless sensor networks. In: Proceedings of of the IEEE military communication conference (MILCOM'05), Atlantic City, NJ, Oct 2005

Cheng B, Chang T (2007) Enhancing ultrasonic imaging with low transient pulse shaping. IEEE Trans Ultrason Ferroelectr Freq Control 54(3):627–635

Clarke DW (1995) Sensor, actuator, and loop validation. IEEE Control Syst Mag 15:39–45

Coello Coello CA, Lamont GB, Van Veldhuizen DA (2007) Evolutionary algorithms for solving multi-objectives problems. Springer, New York

Cross N R, Dotson J R, Flank D R, Peggs G N, Cox M G, Forbes A B, Corta R, O'Donnell J, Prieto E. (1998) A large reference artefact for CMM verification, National Physical Laboratory, NPL Report CLM 6 May 1998

Crossbow Technology (2010). http://www.xbow.com/. Accessed 15 October 2010

Curtis MA, Farago FT (1994) Handbook of dimensional measurement. Industrial Press Inc, New York

Cuypers W, Van Gestelb N, Voeta A, Kruthb JP, Mingneaua J, Bleysc P (2009) Optical measurement techniques for mobile and large-scale dimensional metrology. Opt Lasers Eng 47(3–4):292–300

Dasgupta K, Kukreja M, Kalpakis K (2003) Topology-aware placement and role assignment for energy-efficient information gathering in sensor networks. In: Proceedings of the 8th IEEE symposium on computers and communication (ISCC'03), Kemer-Antalya, Turkey, June 2003

Daumas M (1989) Scientific instruments of the seventeenth and eighteenth centuries and Their Makers, Portman Books, London

Delepaut C, Vandendrope L, Eugène C (1986) Ultrasonic three-dimensional detection of obstacles for mobile robots. In proceedings of 16th international symposium of industrial robots (ISIR 86 Brussels) pp 483–90

Dhillon SS, Chakrabarty K, Iyengar SS (2002) Sensor placement for grid coverage under imprecise detections. In: Proceedings of the 5th international conference on information fusion, Annapolis, MD, July 2002, pp 1581–1587

Doğançay K (2005) Bearings-only target localization using total least squares. Signal Process 85(9):1695–1710

Eastwood SJ, Webb P, Mckeown C (2003) The use of the TI^2 manufacturing system on a double-curvature aerospace panel. In: Proceedings of institute of mechanical engineers. Part B J Eng Manuf 217(6):849–855

Estler WT, Edmundson KL, Peggs GN, Parker DH (2002) Large scale metrology–an update. CIRP Ann—Manuf Technol 51(2):587–609

FARO EUROPE GmbH Co. KG (2004) New Faro laser tracker SI. 2: tougher with exclusive features Faro UK technical specification sheet

Feng CXJ, Saal AL, Salsbury JG, Ness AR, Lin GCS (2007) Design and analysis of experiments in CMM measurement uncertainty study. Precis Eng 31(2):94–101

Ferri C, Mastrogiacomo L, Faraway J (2010) Sources of variability in the set-up of an Indoor GPS. Int J Comput Integr Manuf 23(6):487–499

Figueroa JF, Lamancusa J (1992) A method for accurate detection of time-of-arrival- analysis and design of an ultrasonic ranging system. J Acoust Soc Am 91(1):486–494

Fischler M, Bolles R (1981) Random sample consensus: a paradigm for model fitting with applications to image analysis and automated cartography. Commun ACM 24(6):381–395

Fischer ML, Price PN, Thatcher TL, Schwalbe CA, Craig MJ, Wood EE, Sextro RG, Gadgil AJ (2001) Rapid measurements and mapping of tracer gas concentrations in a large indoor space. Atmos Environ 35(1):2837–2844

Franceschini F, Galetto M, Settineri L (2002) On-line diagnostic tools for CMM performance. Int J Adv Manuf Technol 19(1):125–130

Franceschini F, Galetto M (2007) A taxonomy of model-based redundancy methods for CMM on-line performance verification. Int J Technol Manag 37(1–2):104–124

Franceschini F, Galetto M, Maisano D, Mastrogiacomo L (2008) The problem of distributed wireless sensors positioning in the mobile spatial coordinate measuring system (MScMS). In: Proceedings of the 9th Biennial ASME conference on engineering systems design and analysis ESDA08, Haifa, Israel, 7–9 July 2008

Franceschini F, Maisano D, Mastrogiacomo L (2009a) Mobile spatial coordinate measuring system (MScMS) and CMMs: a structured comparison. Int J Adv Manuf Technol 42(1):1089–1102

Franceschini F, Galetto M, Maisano D, Mastrogiacomo L (2009b) On-line diagnostics in the mobile spatial coordinate measuring system (MScMS). Precis Eng 33(1):408–417

Franceschini F, Galetto M, Maisano D, Mastrogiacomo L (2009c) A review of localization algorithms for distributed wireless sensor networks in manufacturing. Int J Comput Integr Manuf 22(7):698–716

Franceschini F, Galetto M, Maisano D, Mastrogiacomo L (2009d) Mobile spatial coordinate measuring system (MScMS)—introduction to the system. Int J Prod Res 47(14):3867–3889

Franceschini F, Mastrogiacomo L, Pralio B (2010a) An unmanned aerial vehicle-based system for large scale metrology applications. Int J Prod Res 48(13):3867–3888

Franceschini F, Maisano D, Mastrogiacomo L, Pralio B (2010b) Ultrasound transducers for large-scale metrology: a performance analysis for their use by the MScMS. IEEE Trans Instrum Meas 59(1):110–121

Frank PM (1990) Fault diagnosis in dynamic systems using analytical and knowledge-based redundancy–A survey and some new results. Automatica 26(1):459–474

Froome KD, Bradsell RH (1966) A new method for the measurement of distances up to 5000 ft by means of a modulated light beam. J Sci Instrum 43(1):129–33

Galetto M, Pralio B (2010a) Optimal sensor positioning for large scale metrology applications. Precis Eng 34(3):563–577

Galetto M, Mastrogiacomo L, Pralio B, Spagnolo C (2010b) Indoor environmental mapping by means of autonomous guided agents. In: Proceedings of ASME 2010 10th Biennial conference on engineering systems design and analysis (ESDA2010), Istanbul, Turkey

Galetto M, Mastrogiacomo L, Pralio B (2011) MScMS-II: an innovative IR-based indoor coordinate measuring system for large-scale metrology applications. Int J Adv Manuf Technol 52(1–4):291–302

Gay D, Levis P, Von Behren R, Welsh M, Brewer E, Culler D (2003) The nesC language: a holistic approach to networked embedded systems. In: Proceedings of programming language design and implementation (PLDI), June 2003

Gertler JJ (1998) Fault detection and diagnosis in engineering systems. Marcel Dekker, New York

Ghosh A, Das SK (2008) Coverage and connectivity issues in wireless sensor networks: a survey. Pervasive Mob Comput 4(1):303–334

Gibson J (1999) The mobile communications handbook. IEEE Press

Glover F, Laguna M (1997) Tabu search. Kluwer, Boston

Godfrey LG, Orme CD, Santos Silva JMC (2006) Simulation-based tests for heteroskedasticity in linear regression models: some further results. Econom J 9(1):76–97

Goldberg DE (1989) Genetic algorithms in search, optimization, and machine learning. Addison-Wesley, Reading

Gonzales-Banos H H, Latombe J C (2001) A randomized art-gallery algorithm for sensor placement. In: Proceedings of the 17th ACM symposium on computational Geometry (SoCG'01), Boston, MA, June 2001

Gustafsson F, Gunnarsson F (2003) Positioning using time difference of arrival measurements. In: Proceedings of the IEEE international conference on acoustics, speech, signal processing (ICASSP). Hong Kong 6, pp 553–556

Hartley RI, Sturm P (1997) Triangulation. Comput Vis Image Underst 68:146–157

Hartley RI, Zisserman A (2004) Multiple view geometry in computer vision, Cambridge University Press

Hazas M, Ward A (2002) A novel broadband ultrasonic location system. In: Proceedings of the fourth international conference on Ubiquitous computing (UbiComp 2002), Goteborg, Sweden, 1:264–280

Hedges TM, Takagi H, Pratt T, Sobel MJ (2003) Position measurement system and method using cone math calibration. United States Patent US6535282B2

Heikkilä J, Silvén O (1997) A four-step camera calibration procedure with implicit image correction. In: Proceedings of IEEE computer society conference on computer vision and pattern recognition (CVPR97), San Juan, Puerto Rico, pp 1106–1112

Henry MP, Clarke DW (1992) The self-validating sensor: rationale, definitions and examples. Control Eng Pract 1:585–610

Hexagon Metrology (2010). http://www.hexagonmetrology.com/. Accessed 15 Oct 2010

Hofmann-Wellenhof B, Lichtenegger H, Collins J (2001) GPS: theory and practice. Springer, Wien

Holland JH (1992) Adaptation in natural and artificial systems. MIT Press Cambridge, MA

Isermann R (1984) Process fault detection based on modeling and estimation methods–a survey. Automatica 20(1):387–404

Isermann R (1993) Fault diagnosis of machines via parameter estimation and knowledge processing–tutorial paper. Automatica 29(1):815–835

Ishizuka M, Aida M (2004) Performance study of node placement in sensor networks. In: Proceedings of the 24th international conference on distributed computing systems workshops–W7. Hachioji, Tokyo, Japan

ISO 10012-2 (1997) Quality assurance for measuring equipment - Part 2: guidelines for control of measurement processes

ISO 10360-2 (2001) Geometrical product specifications (GPS)—acceptance and reverification tests for coordinate measuring machines (CMM), Geneva, Switzerland

ISO/TR 16015:2003 (2003) Geometrical product specifications (GPS)—systematic errors and contributions to measurement uncertainty of length measurement due to thermal influences, ISO, Geneva

ISO/TS 15530-6 (Working Draft) (2000) Geometrical product specifications (GPS)—Coordinate measuring machines (CMM): techniques for determining the uncertainty of measurements. Part 6: Uncertainty assessment using un-calibred workpieces, ver. 18 Sept 2000

Jakevicius L, Demcenko A (2008) Ultrasound attenuation dependence on air temperature in closed chambers. Ultragarsas (Ultrasound) 63(1):18–22

JCGM 100:2008 (2008a) Evaluation of measurement data—Guide to the expression of uncertainty in measurement . International Organization for Standardization, Geneva, Switzerland

JCGM 200:2008 (2008b) VIM—International vocabulary of metrology—basic and general concepts and associated terms (VIM). International Organization for Standardization, Geneva, Switzerland

Jiménez AR, Seco F (2005) Precise localisation of archaeological findings with a new ultrasonic 3D positioning sensor. Sens Actuator A 123–124:224–233

Jin Y, Olhofer M and Sendhoff B (2001) Dynamic weighted aggregation for evolutionary multi-objective optimization: why does It works and how?. In: Proceedings of genetic and evolutionary computation conference, pp 1042–1049

Johansson J, Gustafsson M, Delsing J (2006) Ultra-low power transmit/receive ASIC for battery operated ultrasound measurement systems. Sens Actuator 125(1):317–328

Kang S, Tesar D (2004) A noble 6-DOF measurement tool with indoor GPS for metrology and calibration of modular reconfigurable robots. In: Proceedings of IEEE ICM international conference on mechatronics, Istanbul, Turkey

Kayani A, Jamshidi J (2007) Measurement assisted assembly for large volume aircraft wing structures. In: Proceedings of DET2007, 4th international conference on digital enterprise technology, Bath, UK, 19–21 Sept 2007

Kažys R, Mažeika L, Šliteris R, Voleišis A (2007) Online profiling of nonplanar objects by high-resolution air-coupled ultrasonic distance measurements. IEEE Trans Instrum Meas 56(5):1825–1830

Kino GS (1987) Acoustic waves: devices, imaging, and analog signal processing. Prentice-Hall, New Jersey

Kirkpatrick S, Gelatt C, Vecchi M (1983) Optimization by simulated annealing. Science 220:671–680

Koza JR (1992) Genetic programming. On the programming of computers by means of natural selection. The MIT Press, Cambridge

Kuka Roboter (2002) KR 240-2, KR 240 L210-2, KR 240 L180-2 (Serie 2000) Technical data

Kwakkernaat M, De Jong YLC, Bultitude RJC, Herben M (2008) High-resolution angle-of-arrival measurements on physically-nonstationary mobile radio channels. IEEE Trans Antennas Propag 56(8):2720–2729

Laarhoven PJM, Aarts EHL (1987) Simulated annealing: theory and applications. Kluwer, Norwell

Laguna M, Roa JO, Jimenez AR, Seco F (2009) Diversified local search for the optimal layout of beacons in an indoor positioning system. IIE Trans 41:247–259

Lamancusa J, Figueroa JF (1990) Ranging errors caused by angular misalignment between ultrasonic transducer pairs. J Acoust Soc Am 87(3):1327–1335

Leica Geodetic Services Inc. (2010). http://www.geodetic.com. Accessed date: 15/10/2010

Lilienthal AJ, Duckett T (2004) Building gas concentration gridmaps with a mobile robot. Robot Auton Syst 48:3–16

Liu X, Mohapatra P (2005) On the deployment of wireless sensor networks. In: Proceedings of the 3rd international workshop on measurement, modelling, and performance analysis of wireless sensor networks (SenMetrics), San Diego, CA, July 2005, pp 78-85

Longuet-Higgins HC (1981) A computer algorithm for reconstructing a scene from two projections. Nature 293:133–135

Lourakis MIA, Argyros AA (2009) SBA: a software package for generic sparse bundle adjustment. ACM Trans Math Softw 36(1):1–30

Luhmann T, Robson S, Kyle S, Harley I (2006) Close range photogrammetry. Wiley, New York

Magori V (1994) Ultrasonic sensors in air. Proc Ultrason Symp 1:471–481

Magori V, Walker H (1987) Ultrasonic presence sensors with wide range and high local resolution. IEEE Trans Ultrason Ferroelectr Freq Control 34(2):201–211

Mahajan A, Figueroa F (1999) An automatic self-installation and calibration method for a 3D position sensing system using ultrasonics. Robot Auton Syst 28(1):281–294

Maisano D, Jamshidi J, Franceschini F, Maropoulos P, Mastrogiacomo L, Mileham A, Owen G (2007) A review of two multistation large volume measuring instrument: MScMS and iGPS. In: Proceedings of DET2007, 4th international conference on digital enterprise technology. Bath, UK,19-21 Sept 2007, pp 418-425

Maisano D, Jamshidi J, Franceschini F, Maropoulos P, Mastrogiacomo L, Mileham A, Owen G (2008) Indoor GPS: system functionality and initial performance evaluation. Int J Manuf Res 3(3):335–349

Maisano D, Jamshidi J, Franceschini F, Maropoulos PG, Mastrogiacomo L, Mileham AR, Owen GW (2009) Comparison of two distributed large volume measurement systems: MScMS and iGPS, Proceedings of the institution of mechanical engineers. Part B J Eng Manuf 223(5): 511–521

Maisano D, Mastrogiacomo L (2010) An empirical regressive model to improve the metrological performance of mobile spatial coordinate measuring systems proceedings of the institution of mechanical engineers. Part B J Eng Manuf 224(4):663–677

Mallows CL (1973) Some comments on Cp. Technometrics 15(1):661–675

Manthey W, Kroemer N, Magori V (1991) Ultrasonic transducers and transducer arrays for applications in air. Meas Sci Technol 3(3):249–261

Martin JM, Jiménez AR, Seco F, Calderon L, Pons JL, Ceres R (2002) Estimating the 3D-position from time delay data of US-waves: experimental analysis and new processing al-gorithm. Sens Actuators 101(3):311–321

Mason S (1997) Heuristic reasoning strategy for automated sensor placement. Photogramm Eng Remote Sens 63:1093–1102

Mastrogiacomo L, Maisano D (2010) Network localisation procedures for Experimental evaluation of Mobile Spatial coordinate Measuring System (MScMS). Int J Adv Manuf Technol 48(9):859–870

Megerian S, Koushanfar F, Qu G, Veltri G, Potkonjak M (2002) Exposure in wireless sensor networks: theory and practical solutions. Wirel Netw 8(1):443–454

Meguerdichian S, Koushanfar F, Potkonjak M, Srivastava M B (2001) Coverage problems in wireless Ad-Hoc sensor networks. In: Proceedings of the 20th annual joint conference of the IEEE computer and communications societies (INFOCOM), Anchorage, AK, April 2001, pp. 1380–1387

Metronor Corporate (2010). http://www.metronor.com. Accessed date: 15/10/2010

Mikhail EM, Bethel JS, McGlone JC (2001) Introduction to modern photogrammetry. Wiley, New York

MIT Computer Science and Artificial Intelligence Lab (2004) Cricket v2 User Manual

Montgomery DC (2005) Introduction to statistical process control. J. Wiley, New York

Montgomery DC (2008) Design and Analysis of Experiments, 7th Edn. Wiley, New York

Moore D, Leonard J, Rus D, Teller SS (2004) Robust distributed network localization with noisy range measurements. In: Proceedings of SenSys 2004, November, Baltimore, MD, 50–6

Muelaner JE, Wang Z, Jamshidi J, Maropoulos PG, Mileham AR, Hughes EB, Forbes AB (2008a) iGPS—An initial assessment of technical and deployment capability. In: Proceedings of the 3rd international conference on manufacturing engineering (ICMEN), Chalkidiki, Greece, 2008

Muelaner J, Hughes B, Forbes A, Maropoulos P, Jamshidi J, Wang Z, Sun W (2008b) iGPS capability assessment. Large volume metrology conference, Liverpool

Nagpal R, Shrobe H, Bachrach J (2003) Organizing a global coordinate system from local information on an Ad-Hoc sensor network. In: Proceedings of the international workshop on information processing in sensor networks (IPSN 2003), Palo Alto, CA

Nasipuri A, Li K (2002) A directionality based location discovery scheme for wireless sensor networks. In: Proceedings of ACM international workshop on wireless sensor networks and applications (WSNA'02), pp 105–111

NaturalPoint Inc (2010). http://www.naturalpoint.com/optitrack. Accessed date: 15/10/2010

Navarro L, Dolan J, Khosla P (2004) Optimal sensor placement for cooperative distributed vision. In: Proceedings of the international conference on robotics and automation (ICRA'04), New Orleans, LA, April 2004

Niculescu D, Nath B (2001) Ad-Hoc positioning system (APS). In: Proceedings of IEEE global communications conference (GlobeCom'01), pp 2926–2931

Niculescu D, Nath B (2003) Ad-Hoc positioning system (APS) using AOA. In: Proceedings of IEEE annual joint conference IEEE computer and communications societies (INFOCOM'03), pp 1734–1743

Nikon Metrology (2010). http://www.nikonmetrology.com. Accessed date: 15/10/2010

Northern Digital Inc. (2010). http://www.ndigital.com/industrial/products-pcmm.php. Accessed date: 15/10/2010

Oh SC, Tan CH, Kong FW, Tan YS, Ng KH, Ng GW, Tai K (2007) Multiobjective optimization of sensor network deployment by a genetic algorithm. In: Proceedings of IEEE congress on evolutionary computation, IEEE, Piscataway, NJ, pp 3917–3921

Olague G, Mohr R (2002) Optimal camera placement for accurate reconstruction. Pattern Recognit 35(1):927–944

Osyczka A (2002) Evolutionary algorithms for single and multicriteria design optimization. Physica, Verlag

Overmars MH (1997) Designing the computational geometry algorithms library CGAL. Applied computational geometry towards geometric engineering. Springer-Verlag, Berlin

Padula S L, Kincaid R K (1999) Optimization strategies for sensor and actuator placement. NASA/TM-1999-209126

Pan J, Cai L, Hou YT, Shi Y, Shen SX (2005) Optimal base-station locations in two-tiered wireless sensor networks. IEEE Trans Mobile Comput 4:458–473

Parrilla M, Anaya JJ, Fritsch C (1991) Digital signal processing techniques for high accuracy ultrasonic range measurements. IEEE Trans Instrum Meas 40(4):759–764

Patwari N, Ash J, Kyperountas S, Hero A III, Moses R, Correal N (2005) Locating the nodes–cooperative localization in wireless sensor networks. IEEE Signal process Mag 22(4):54–69

Peggs GN, Maropoulos PG, Hughes EB, Forbes AB, Robson S, Ziebart M, Muralikrishnan B (2009) Recent developments in large-scale dimensional metrology, proceedings of the institution of mechanical engineers. Part B J Eng Manuf 223(6):571–595

Petrushin V A, Wei G, Shakil O, Roqueiro D, Gershman V (2006) Multiple-sensor indoor surveillance system. In: Proceedings of the 3rd Canadian conference on computer and robot vision (CRV'06), Quebec City, June 2006

Phillips SD, Sawyera D, Borchardta B, Warda D, Beutelb DE (2001) A novel artifact for testing large coordinate measuring machines. J Int Soc Precis Eng Nanotechnol 25:29–34

Piontek H, Seyffer M, Kaiser J (2007) Improving the accuracy of ultrasound-based localisation systems. Pers Ubiquitous Comput 11(6):439–449

Piratelli-Filho A, Di Giacomo B (2003) CMM uncertainty analysis with factorial design. Precis Eng 27(3):283–288

Poduri S, Pattem S, Krishnamachari B, Sukhatme G S (2006) Sensor network configuration and the curse of dimensionality. In: Proceedings of the 3rd IEEE workshop on embedded networked sensors, Cambridge, MA, May 2006

Pompili D, Melodia T, Akyildiz (2006) Deployment analysis in underwater acoustic wireless sensor networks.In: Proceedings of the ACM International Workshop on Underwater Networks (WUWNet), Los Angeles, Sept 2006

Priyantha N B, Chakraborty A, Balakrishnan H (2000) The cricket location-support system. In: Proceedings 6th ACM MOBICOM, Boston, MA, pp 32–43

Priyantha NB, Balakrishnan H, Demaine E, Teller S (2003) Anchor-free distributed localization in sensor networks, Tech. Rep. 892, MIT Lab. for Comp. Sci

Priyantha N B, Balakrishnan H, Demaine H D, Teller S (2005) Mobile-Assisted Localisation in Wireless Sensor Networks. In: Proceedings of the 24th annual joint conference of the IEEE communications society on computer communications (INFOCOM 2005), vol 1. Miami, Florida, 13–17 March, pp 172–183

Puttock MJ (1978) Large-scale metrology. Ann CIRP 21(1):351–356

Ramadan R, El-Rewini H, Abdelghany K (2007) Optimal and approximate approaches for deployment of heterogeneous sensing devices. EURASIP J Wirel Commun Netw 2007 Article ID 54731

Ray PK, Mahajan A (2002) A genetic algorithm-based approach to calculate the optimal configuration of ultrasonic sensors in a 3D position estimation system. Robotics Auton Syst 41:161–177

Roa JO, Jimenez AR, Seco F, Prieto JC, Ealo J (2007) Optimal placement of sensors for trilateration: regular lattices vs meta-heuristic solutions. Lecture notes in computer science. Comput Aided Syst Theory EUROCAST 4739(2007):780–787

Ronchetti E, Staudte RG (1994) A robust version of Mallow's Cp. J Am Stat Assoc 89:550–559

Rooks B (2001) Automatic wing box assembly developments. Ind Robot 28(4):297–301

Safigianni AS, Tsompanidou CG (2005) Measurements of electric and magnetic fields due to the operation of indoor power distribution substations. IEEE Trans Power Deliv 20:1800–1805

Sandwith S, Predmore R (2001) Real-time 5-micron uncertainty with laser tracking interferometer systems using weighted trilateration. In: Proceedings of 2001 boeing large scale metrology seminar. St. Louis, MO

Savio E, Hansen H N, De Chiffre L (2002) Approaches to the calibration of freeform artifacts on coordinate measuring machines. Annals of CIRP 51/1, San Sebastian, Spain, pp 433–436

Savvides A, Han CC, Srivastava M (2001) Dynamic fine-grained localisation in Ad-Hoc networks of sensors. In: Proceedings of the 7th IEEE/ACM annual conference on mobile computing and networks (MobiCOM'01), Rome, Italy, July 2001

Schwager M, McLurkin J, Slotine JJE, Rus D (2008a) From theory to practice: distributed coverage control experiments with groups of robots. In: Proceedings international symposium on experimental robotics, Athens, July 2008

Schwenke H, Knapp W, Haitjema H, Weckenmann A, Schmitt R, Delbressine F (2008b) Geometric error measurement and compensation of machines—an update. CIRP Annals—Manuf Technol 57:660–675

Sinriech D, Shoval S (2000) Landmark configuration for absolute positioning of autonomous vehicles. IIE Trans 32:613–624

Svoboda T, Martinec D, Pajdla T (2005) A convenient multi-camera self-calibration for virtual environments. Presence: Teleoperators Virtual Environ 14(4):407–422

Syswerda G (1989) Uniform crossover in genetic algorithms. In: Proceedings of the 3rd International conference on genetic algorithms. San Mateo, CA

Tang J, Hao B, Sen A (2006) Relay node placement in large scale wireless sensor networks. Comput Commun, Special issue Wirel Sens Netw 29:490–501

TinyOS (2010). http://www.tinyos.net. Accessed 15 October 2010

Toda M (2002) Cylindrical PVDF film transmitters and receivers for air ultrasound. IEEE trans ultrason ferroelectr freq control 49(5):626–634

Toda M, Dahl J (2005) PVDF corrugated transducer for ultrasonic ranging sensor. Sens Actuators 134:427–435

Tong CC, Figueroa JF, Barbieri E (2001) A method for short or long range time-of-flight measurements using phase-detection with an analog circuit. IEEE Trans Instrum Meas 50(5):1324–1328

Triggs B, McLauchlan P F, Hartley R I, Fitzgibbon AW (2000) Bundle adjustment—A modern synthesis. London, UK: Springer-Verlag, 298–372

Van Veldhuizen DA, Lamont GB (2000) Multiobjective evolutionary algorithms: analyzing the state-of-the-art. Evolutionary Comput 7:1–26

VDI/VDE Guideline 2634 Part 1 (2002) Optical 3-D measuring systems and imaging systems with point-by-point probing, Beuth Verlag, Berlin

Vicon Inc (2010). http://www.vicon.com. Accessed 15 October 2010

Wang XR, Xing GL, Zhang YF, Lu CY, Pless R, Gill C (2003) Integrated coverage and connectivity configuration in wireless sensor networks. In: Proceedings of the 1st international conference on embedded networked sensor system, 5–7 Nov, Los Angeles, CA, pp 28–39

Wang G, Cao G, La Porta T (2004a) Movement-assisted sensor deployment. In: Proceedings of the 23rd international annual joint conference of the IEEE computer and communications societies, Hong Kong, March 2004

Wang G, Cao G, La Porta T (2004b) Proxy-based sensor deployment for mobile sensor networks. In: Proceedings of the 1st international conference on mobile Ad-Hoc and sensor systems (MASS'04), Fort Lauderdale, Florida, Oct 2004

Wang Q, Xu K, Takahara G, Hassanein H (2006) Deployment for information oriented sensing coverage in wireless sensor networks. In: Proceedings of the 49th IEEE global telecommunication conference (Globecom'06), San Francisco, CA, Nov 2006

Wang Z, Mastrogiacomo L, Maropoulos P G, Franceschini F (2009) Experimental testing of the dynamic tracking performance of iGPS and laser tracker. In: Proceedings of laser metrology and machine performance IX, LAMDAMAP 2009, Teddin (UK) 30th June–2nd July 2009, pp 305–314

Webb P, Eastwood SJ (2004) An evaluation of the TI^2 manufacturing system for the machining of airframe subassemblies. In: Proceedings of Institute of Mechanical Engineers. Part B: J Eng Manuf 218(7):819–826

Welch G, Bishop G, Vicci L, Brumback S, Keller K (2001) High-performance wide-area optical tracking the hiball tracking system. Presence: Teleoperators Virtual Environ 10(1):1–21

Wolf P, Ghilani C (1997) Adjustment computations: statistics and least squares in surveying and GIS. Wiley, New York

Wu H, Wang C, Tzeng N (2008) Novel self-configurable positioning technique for multihop wireless networks. IEEE/ACM Trans Netw 13:609–621

Xu K, Hassanein H, Takahara G, Wang W (2005a) Relay node deployment strategies in heterogeneous wireless sensor networks: single-hop communication case. In: Proceedings of the IEEE global telecommunication conference. St. Louis, MO

Xu K, Hassanein H, Takahara G Wang W (2005b) Relay node deployment strategies in heterogeneous wireless sensor networks: multiple-hop communication case. In: Proceedings of the 2nd IEEE conference on sensor and Ad-Hoc communications and networks, Santa Clara, CA

Xu K, Wang Q, Hassanein H S, Takahara G (2005c) Optimal wireless sensor networks (WSNs) deployment: minimum cost with lifetime constraints. In: Proceedings of the IEEE international conference on wireless and mobile computing, networking and communications (WiMob05), Montreal, Canada, Aug 2005

Xu K, Takahara G, Hassanein H (2006) On the robustness of grid-based deployment in wireless sensor networks. In: Proceedings of international wireless communications and mobile computing conference (IWCMC), Vancouver, BC, Canada, pp 1183–1188

Yang B, Scheuing J (2005) Cramer-Rao bound and optimum sensor array for source localization from time differences of arrival. In: Proceedings of IEEE international conference on acoustics, speech, and signal processing, IEEE, Piscataway, NJ, pp 961–964

Yarlagadda R, Al-Dhahir IAN, Hershey J (2000) GPS GDOP Metric. IEEE Proc-Radar, Sonar Navig 147:259–264

Younis M, Akkaya K (2008) Strategies and techniques for node placement in wireless sensor networks: a survey. Ad-Hoc Netw 6:621–655

Zakrzewski J (2003) Error and uncertainty reduction–challenge for a measuring systems designer. Meas Sci Rev 3(1):31–34

Zhang X, Wicker S B (2004) How to distribute sensors in a random field? In: Proceedings of the 3rd international symposium on information processing in sensor networks (IPSN '04), Berkeley, CA, April 2004

Zhang D, Rolt S, Maropoulos PG (2005) Modelling and optimization of novel laser multilateration schemes for high-precision applications. Meas Sci Technol 16:2541–2547

Index